Your Quick & Easy Path to your FIRST Ham Radio License!

By Ward Silver, NØAX

Sixth Edition

Contributing Editor:
Mark Wilson, K1RO

Editorial Assistants
Maty Weinberg, KB1EIB

Production Staff:
David Pingree, N1NAS, Senior Technical Illustrator
Jodi Morin, KA1JPA, Assistant Production Supervisor: Layout
Sue Fagan, KB1OKW, Graphic Design Supervisor: Cover Design
Michelle Bloom, WB1ENT, Production Supervisor: Layout

Front Cover Photo: Robert Wood, W5AJ, operates using Morse code during the Midland (Texas) Amateur Radio Club Field Day event. Every June Field Day brings Amateur Radio operators – old and new – together for a weekend of enjoyment on the air. By earning your license, you'll be able to take part in exciting activities like these, and many more! (Photo courtesy of Alan Sewell, N5NA)

ARRL The national association for AMATEUR RADIO®
225 Main Street, Newington, CT 06111-1494
www.arrl.org

Printed in USA

IBSN: 978-1-62595-017-8

Sixth Edition
Second Printing

This book may be used for Technician license exams given beginning July 1, 2014. *QST* and the ARRL website (**www.arrl.org**) will have news about any rules changes affecting the Technician class license or any of the material in this book.

We strive to produce books without errors. Sometimes mistakes do occur, however. When we become aware of problems in our books (other than obvious typographical errors), we post corrections on the ARRL website. If you think you have found an error, please check **www.arrl.org** for corrections and supplemental material. If you don't find a correction there, please let us know by sending e-mail to **pubsfdbk@arrl.org**.

CONTENTS

FOREWORD

Congratulations on beginning your journey to join the diverse group of individuals who make up Amateur Radio! It's a big family — there are more than 700,000 amateurs or "hams" in the United States and more than 3,000,000 around the world. This group is made up from a wide variety of people of all ages and walks of life, all around the world. These people use Amateur Radio to communicate directly, for the most part without relying on any commercial or government systems.

Your interest in ham radio may come from a desire to provide public service to your community, which can range from disaster response to monitoring severe weather to helping with parades or charity events. Perhaps you are one of the growing "do it yourself" community and are interested in learning more about electronics and wireless communication. Or you may have scientific interests that involve radio and similar natural phenomena. And maybe you like the idea of Amateur Radio providing an enjoyable hobby and a means of staying in touch with family and friends independent of the Internet or telephone networks. All are welcome and all are important to Amateur Radio's future.

Recognizing that answering the exam questions correctly is only the beginning, be sure to take advantage of the additional material provided by the ARRL to help you get your license and then learn how to communicate as an amateur operator. The following Introduction section will present the ARRL's website and other online resources to help you. This is a key element of the ARRL's mission to help amateurs — our motto is "Of, By, and For the Amateur." By providing these resources, you will be better prepared to get on the air, have more fun, and be more effective on the air.

Most of the active amateur operators in the United States are ARRL members because they recognize the value of having an active national organization to represent them, provide training and educational opportunities, sponsor activities, and distribute information about the Amateur service. By joining, you support all of these and more. This book is just one of the ARRL's many publications for all levels and interests in Amateur Radio. You don't need a license to join the ARRL — just be interested in Amateur Radio and the ARRL will be interested in you. It's that simple! Welcome to Amateur Radio.

David Sumner, K1ZZ
Chief Executive Officer
Newington, Connecticut
March 2014

When to Expect New Books

A Question Pool Committee (QPC) consisting of representatives from the various Volunteer Examiner Coordinators (VECs) prepares the license question pools. The QPC establishes a schedule for revising and implementing new Question Pools. The current Question Pool revision schedule is as follows:

Question Pool	Current Study Guides	Valid Through
Technician (Element 2)	*The ARRL Ham Radio License Manual*, 3rd Edition *ARRL's Tech Q&A*, 6th Edition	**June 30, 2018**
General (Element 3)	*The ARRL General Class License Manual*, 7th edition *ARRL's General Q&A*, 4th Edition	**June 30, 2015**
Amateur Extra (Element 4)	*The ARRL Extra Class License Manual*, 10th Edition *ARRL's Extra Q&A*, 3rd Edition	**June 30, 2016**

As new Question Pools are released, ARRL will produce new study materials before the effective date of the new pools. Until then, the current Question Pools will remain in use, and current ARRL study materials, including this book, will help you prepare for your exam.

As the new Question Pool schedules are confirmed, the information will be published in *QST* and on the ARRL website at **www.arrl.org**.

Online Review and Practice Exams
ARRL Exam Review for Ham Radio

Use this book with ARRL Exam Review for Ham Radio to take randomly-generated practice exams using questions from the actual examination question pool. You won't have any surprises on exam day Go to **www.arrl.org/examreview**

Introduction

The Technician License

Earning a Technician Amateur Radio license begins your enjoyment of ham radio. Topics covered by the exam provide you with a good introduction to basic radio requiring only modest math skills. You are sure to find the operating privileges available to a Technician licensee to be worth the time spent learning about Amateur Radio. After passing the exam, you will be able to operate on every frequency above 50 megahertz available to the Amateur service. You also gain privileges on the traditional "shortwave" 80, 40, 15 and 10 meter amateur bands. With this broad set of operating privileges, you'll be ready to experience the excitement of Amateur Radio!

Perhaps your interest is in Amateur Radio's long history of public service, such as providing emergency communications in time of need. You might have experience with computer networks leading you to explore the digital mode technology used in ham radio. If your eyes turn to the stars on a clear night, you might enjoy tracking the amateur satellites and using them to relay your signals to other amateurs around the world! Your whole family can enjoy Amateur Radio, taking part in outdoor activities such as ARRL Field Day and mobile operating during a vacation or weekend drive.

An Overview of Amateur Radio

Earning an Amateur Radio license is a special achievement. The more than 700,000 people in the US who call themselves Amateur Radio operators, or hams, are part of a global family. Radio signals do not know territorial boundaries, so hams have a unique ability to enhance international goodwill. Hams become ambassadors of their country every time they put their stations on the air.

Radio amateurs provide a voluntary, noncommercial, communication service, without any type of payment except the personal satisfaction they feel from a job well done! This is especially true during natural disasters or other emergencies when the normal lines of communication are out of service. In the aftermath of hurricanes such as Katrina (2005) and Sandy (2012), and 2013's devastating floods in Bangkok and typhoon Haiyan, hams traveled to the stricken areas to establish communications links until normal systems were restored. Thousands more relayed information around the country and the world. In every US county and city, organized groups of amateur operators train and prepare to support their communities during disasters and emergencies of every type.

Hams have made many important contributions to the field of electronics and communications since Amateur Radio's beginnings a century ago and this tradition continues today. Today, hams relay signals through their own satellites, bounce signals off the moon, relay messages automatically through computerized radio networks and use any number of other "exotic" communications techniques.

Amateur Radio experimentation is yet another reason many people become part of this self-disciplined group of trained operators, technicians and electronics experts — an asset to any country. Amateurs talk from hand-held transceivers through mountaintop repeater stations that can relay their signals to other hams' cars or homes or through the Internet around the world. Hams establish wireless data networks, send their own television signals, and talk with other hams around the world by voice. Keeping alive a distinctive traditional skill, they also tap out messages in Morse code.

Who Can Be a Ham…and How?

The US government, through the Federal Communications Commission (FCC), grants all US Amateur Radio licenses. Because of Amateur Radio's tremendously flexible and self-organizing nature, amateurs are expected to know more about their equipment and operating techniques. Unlike other radio services, amateurs organize their own methods of communication, are encouraged to build and repair their own equipment, perform experiments with antennas and with radio propagation, and invent their own protocols and networks. The FCC licensing process ensures that amateurs have the necessary operating skill and electronics know-how to use that flexibility wisely and not interfere with other radio services.

The FCC doesn't care how old you are or whether you're a US citizen. If you pass the examination, the Commission will issue you an amateur license. Any person (except the agent of a foreign government) may take the exam and, if successful, receive an amateur license. It's important to understand that if a citizen of a foreign country receives an amateur license in this manner, he or she is a US Amateur Radio operator. (This should not be confused with a reciprocal operating permit which is covered in the questions of subelement T1.)

License Structure

Anyone earning a new Amateur Radio license can earn one of three license classes — Technician, General and Amateur Extra. Higher-class licenses have more comprehensive examinations. In return for passing a more difficult exam you earn more frequency privileges (frequency space in the radio spectrum). The vast majority of beginners earn the most basic license, the Technician, before beginning to study for the other licenses.

Table 1 lists the amateur license classes you can earn, along with a brief description of their exam requirements and operating privileges. A Technician license gives you the freedom to develop operating and technical skills through on-the-air experience. These skills will help you upgrade to a higher class of license and obtain additional privileges.

The Technician exam, called Element 2, covers some basic radio fundamentals and knowledge of some of the rules and regulations in Part 97 of the FCC Rules. With a little study you'll soon be ready to pass the exam.

Each step up the Amateur Radio license ladder requires the applicant to have passed the lower exams. So if you want to start out as a General class or even an Amateur Extra class licensee, you must first have passed the Technician written exam. You retain credit for all the exam elements of any license class you hold. For example, if you hold a Technician license, you will only have to pass the Element 3 General class written exam to obtain a General class license.

Although there are also other amateur license classes, the FCC is no longer issuing new licenses for them. The Novice license was long considered the begin-

Table 1
Amateur Operator Licenses

License Class	*Written Exam*	*Privileges*
Technician	Basic theory and regulations (Element 2)	All above 50 MHz and limited HF privileges
General	Basic theory and regulations; General theory and regulations (Elements 2 and 3)	All except those reserved for Advanced and Amateur Extra
Amateur Extra	All lower exam elements, plus Amateur Extra theory (Elements 2, 3 and 4)	All amateur privileges

ner's license. Exams for this license were discontinued as of April 15, 2000. The FCC also stopped issuing new Advanced class licenses on that date. They will continue to renew previously issued licenses, however, so you will probably meet some Novice and Advanced class licensees on the air.

As a Technician, you can use a wide range of frequency bands — all amateur bands above 50 MHz (megahertz), in fact. (See **Table 2** and **Figure 1**.) You'll be able to use point-to-point or repeater communications on VHF, use packet radio and other digital modes and networks, even access orbiting satellites or bounce a signal off meteor trails and the Moon! You can use your operating skills to provide public service through emergency communications and message handling.

Station Call Signs

Many years ago, by international agreement, the nations of the world decided to allocate certain call sign prefixes to each country. This means that if you hear a radio station call sign beginning with W or K, for example, you know the station is licensed by the United States. A call sign beginning with the letter G is licensed by Great Britain, and a call sign beginning with VE is from Canada. (All of the amateur call sign prefixes are listed in a table on the ARRL's website, **www.arrl.org**.)

The International Telecommunication Union (ITU) radio regulations outline the basic principles used in forming amateur call signs. According to these regulations, an amateur call sign must be made up of one or two characters (the first one may be a numeral) as a prefix, followed by a numeral, and then a suffix of not more than three letters. The prefixes W, K, N and A are used in the United States. When the letter A is used in a US amateur call sign, it will always be with a two-letter prefix, AA to AL. The continental US is divided into 10 Amateur Radio call districts (sometimes called call areas), numbered 0 through 9. **Figure 2** is a map showing the US call districts.

You may keep the same call sign when you change license class, if you wish. You must indicate that you want to receive a new call sign when you apply for the exam or change your address.

Table 2

US Amateur Bands

Effective Date March 5, 2012

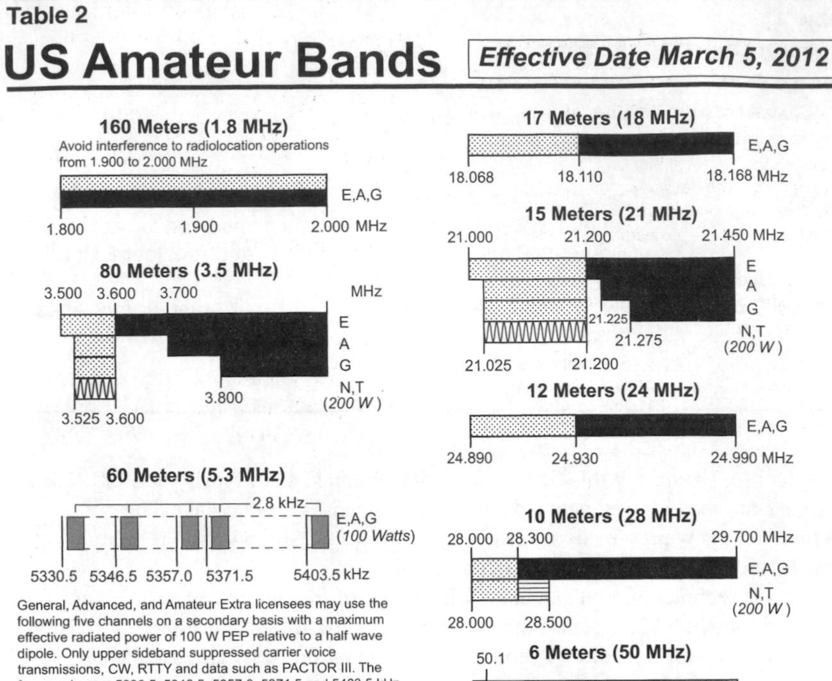

General, Advanced, and Amateur Extra licensees may use the following five channels on a secondary basis with a maximum effective radiated power of 100 W PEP relative to a half wave dipole. Only upper sideband suppressed carrier voice transmissions, CW, RTTY and data such as PACTOR III. The frequencies are 5330.5, 5346.5, 5357.0, 5371.5 and 5403.5 kHz. The occupied bandwidth is limited to 2.8 kHz centered on 5332, 5348, 5358.5, 5373, and 5405 kHz respectively.

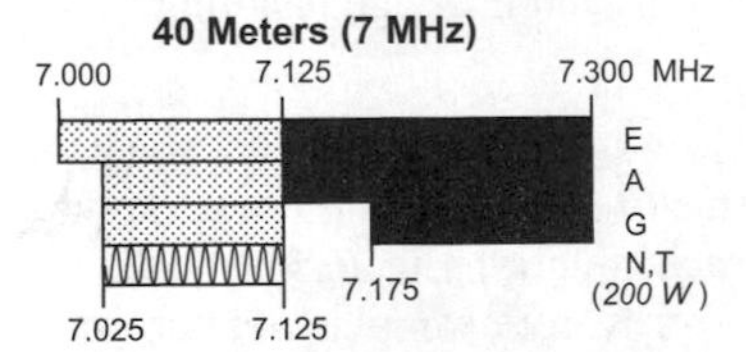

Phone and Image modes are permitted between 7.075 and 7.100 MHz for FCC licensed stations in ITU Regions 1 and 3 and by FCC licensed stations in ITU Region 2 West of 130 degrees West longitude or South of 20 degrees North latitude. See Sections 97.305(c) and 97.307(f)(11). Novice and Technician Plus licensees outside ITU Region 2 may use CW only between 7.025 and 7.075 MHz. See Section 97.301(e). These exemptions do not apply to stations in the continental US.

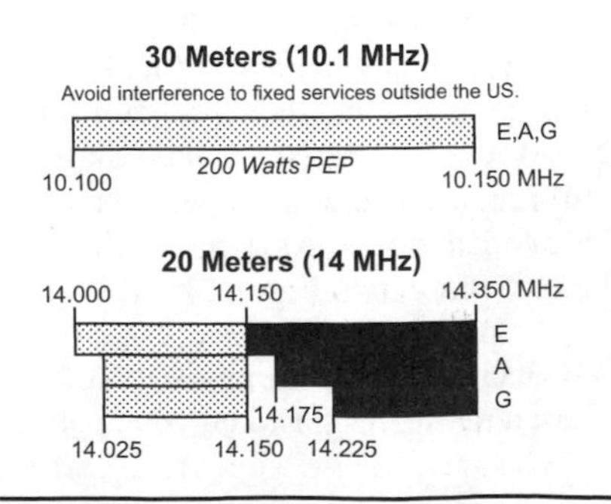

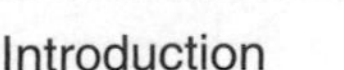

2 Meters (144 MHz)

144.1

E,A,G,T

144.0 148.0 MHz

1.25 Meters (222 MHz)

E,A,G,T

N (25 Watts)

219.0 220.0

222.0 225.0 MHz

*Geographical and power restrictions may apply to all bands above 420 MHz. See The ARRL Operating Manual for information about your area.

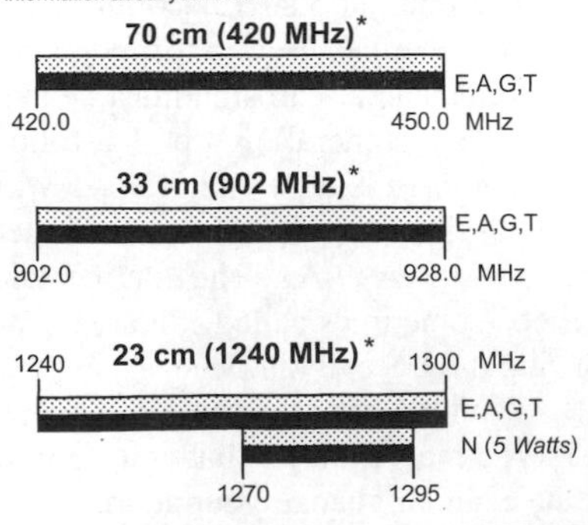

ARRL *The national association for* **AMATEUR RADIO®**

US AMATEUR POWER LIMITS

FCC 97.313 An amateur station must use the minimum transmitter power necessary to carry out the desired communications. **(b)** No station may transmit with a transmitter power exceeding **1.5 kW PEP.**

All licensees except Novices are authorized all modes on the following frequencies:

2300-2310 MHz	47.0-47.2 GHz
2390-2450 MHz	76.0-81.0 GHz
3300-3500 MHz	122.25-123.0 GHz
5650-5925 MHz	134-141 GHz
10.0-10.5 GHz	241-250 GHz
24.0-24.25 GHz	All above 275 GHz

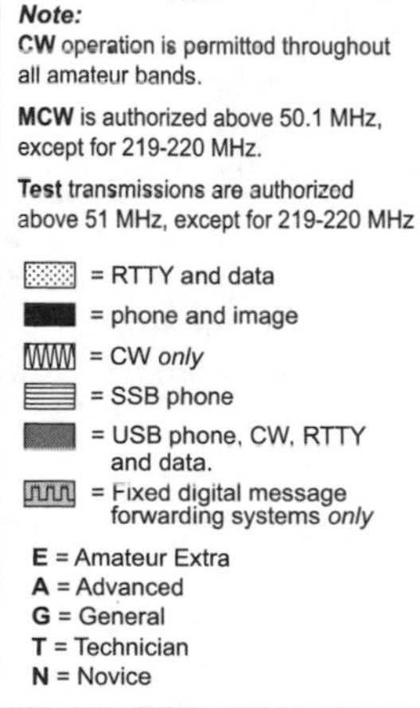

See *ARRLWeb at www.arrl.org* for more detailed band plans.

ARRL ***We're At Your Service***

ARRL Headquarters
225 Main Street, Newington, CT 0611-1494
www.arrl.org
860-594-0200 (Fax 860-594-0259)
email: hq@arrl.org

Publication Orders
www.arrl.org/catalog
Toll-Free 1-888-277-5289 (860-594-0355)
email: orders@arrl.org

Membership/Circulation Desk
Toll-Free 1-888-277-5289 (860-594-0338)
email: membership@arrl.org

Getting Started in Amateur Radio
Toll-Free 1-800-326-3942 (860-594-0355)
email: Newham@arrl.org

Exams 860-594-0300 vec@arrl.org

QnAbands4 rev. 2/13/2012

The FCC also has a vanity call sign system. Under this system the FCC will issue an available call sign selected from a list of your preferred call signs. While there is no fee for an Amateur Radio license, there is a fee for the selection of a vanity call sign. The current fee and details of the vanity call sign system are available on the ARRL website at **www.arrl.org**.

How to Use This Book

The Element 2 exam consists of 35 questions taken from a pool of more than 350 questions. The *ARRL's Tech Q&A* is designed to help you learn about every question in the Technician exam question pool. Every question is presented just as it is in the question pool and as you will encounter it on the exam. Following every question is a short explanation of the answer.

Each chapter of the book covers one subelement from the question pool, beginning with the FCC Rules and ending with Electrical and RF Safety. You may study the questions from beginning to end or select topics in an order that appeals to you.

If you are new to radio, you will probably find it easier to begin with the questions in subelement T3, T4 and T7 to learn about amateur equipment and the basics of radio signals. Then you can move on to the more technical topics covered by T5, T6, T8 and T9. Once you've learned about how radios work, the subelements on operating (T2) and FCC Rules (T1) will make more sense. Finish up with T0 — Electrical and RF Safety — and you'll be ready for your exam!

The *ARRL Ham Radio License Manual* is a good reference companion to the *Tech Q&A*. At the end of the explanation for every question, there

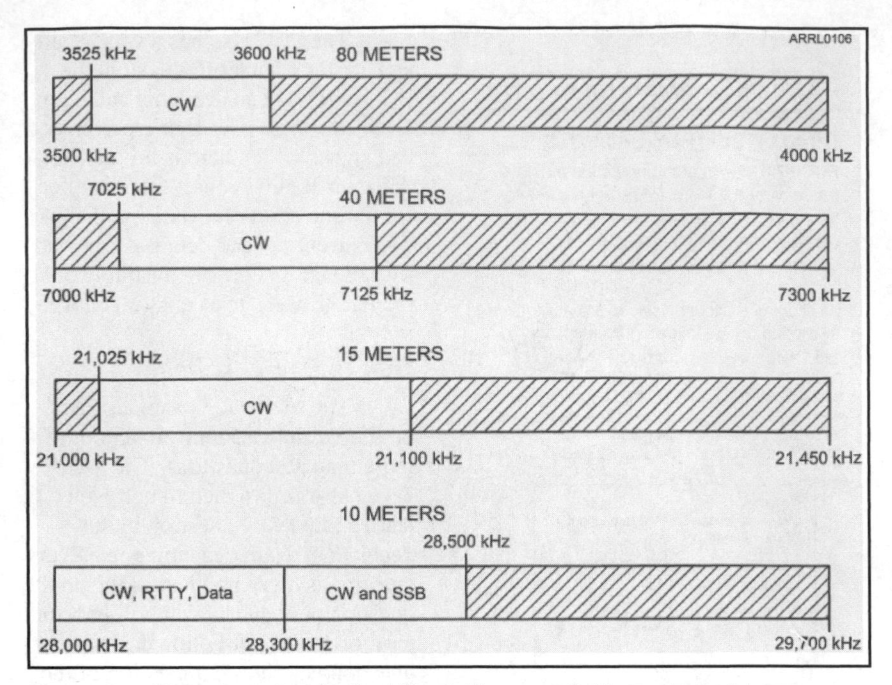

Figure 1 — This chart details the HF priviliges available to Technician licensees.

is a reference to the page in the *Ham Radio License Manual* where you can find a discussion of the topics associated with the question.

There is additional supplemental material on the ARRL's website **www.arrl.org/ham-radio-license-manual** if you need extra help. In particular, there are links to math tutorials and every math problem on the exam is completely worked out to show you how it's done. To make the best use of the online reference material, bookmark the *Ham Radio License Manual* website to use as an on-line reference while you study.

The ARRL's New Ham Desk can answer questions emailed to **newham@arrl.org**. Your question may be answered directly or you might be directed to more instruction material. The New Ham Desk can also help you find a local ham to answer questions. Studying with a friend makes learning the material more fun as you help each other over the rough spots and you'll have someone to celebrate with after passing the exam!

Earning a License

All US amateur exams are administered by Volunteer Examiners who are certified by a Volunteer Examiner Coordinator (VEC) that processes the examination paperwork and license applications for the FCC. A Question Pool Committee selected by the Volunteer Examiner Coordinators maintains the question pools for all amateur exams.

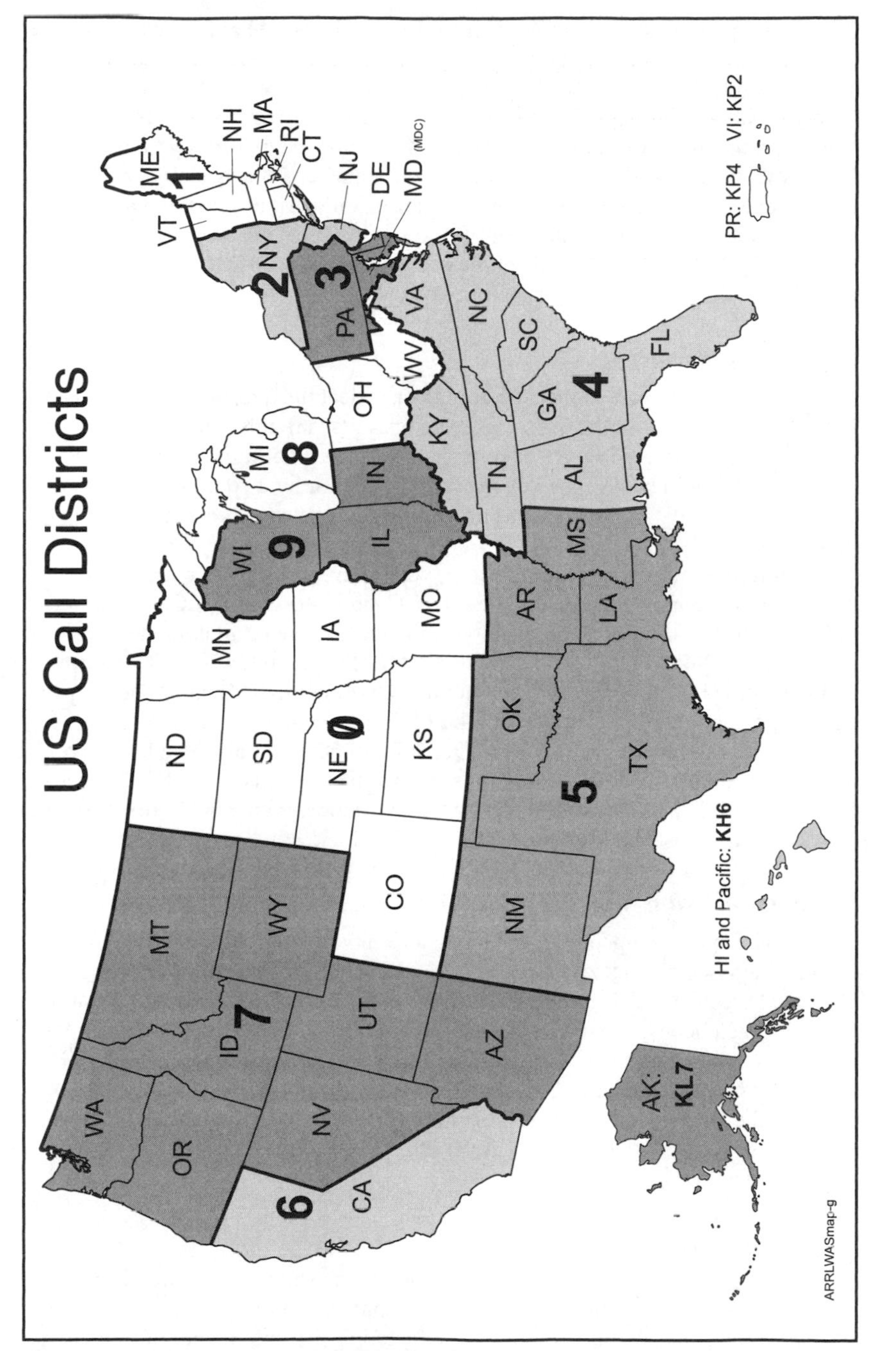

Figure 2—There are 10 US call sign areas. Hawaii and all Pacific possessions are part of the sixth call area and Alaska is part of the seventh. Puerto Rico, the US Virgin Islands and all Caribbean possessions are part of the fourth district.

Once you make the commitment to study and learn what it takes to pass the exam, you will accomplish your goal. Many people pass the exam on their first try, so if you study the material and are prepared, chances are good that you will soon have your license. It may take you more than one attempt to pass the Technician license exam, but that's okay. There is no limit to how many times you can take it. Many Volunteer Examiner teams have several exam versions available, so you may even be able to try the exam again at the same exam session. Time and available exam versions may limit the number of times you can try the exam at a single exam session. If you don't pass after a couple of tries you will certainly benefit from more study of the question pools before you try again.

License Examinations

The FCC allows Volunteer Examiners to select the questions for an amateur exam, but they must use the questions exactly as they are released by the VEC that coordinates the test session. If you attend a test session coordinated by the ARRL/VEC, your test will be designed by the ARRL/VEC or by a computer program created by the VEC. The questions and answers will be exactly as they are printed in this book.

Before you can take an FCC exam, you'll have to fill out a copy of the National Conference of Volunteer Examiner Coordinators (NCVEC) Quick Form 605. This form is used as an application for a new license or an upgraded license. The NCVEC Quick Form 605 is only used at license exam sessions. This form includes some information that the Volunteer Examiner Coordinator's office will need to process your application with the FCC. See **Figure 3**.

You should not use an NCVEC Quick Form 605 to apply for a license renewal or modification with the FCC. Never mail these forms to the FCC, because that will result in a rejection of the application. Likewise, an FCC Form 605 can't be used for a license exam application.

Finding an Exam Session

You can locate upcoming exam sessions in your area by using the ARRL's online Exam Search page. Browse to the ARRL's home page, **www.arrl.org**, and click the "Licensing, Education & Training" button to find complete information about taking a licensing exam. Registration deadlines and the time and location of the exams, are mentioned prominently in publicity releases about upcoming sessions. You can also contact the ARRL/VEC office directly or watch for announcements in the Hamfest Calendar and Coming Conventions columns in *QST*. Many local clubs sponsor exams, so they are another good source of information on exam opportunities.

Taking the Exam

By the time examination day rolls around, you should have already prepared yourself. This means getting your schedule, supplies and mental attitude ready. Plan your schedule so you'll get to the examination site with plenty of time to spare. There's no harm in being early. In fact, you might have time to meet and talk with another applicant which is a great way to calm pretest nerves. Try not to discuss the material that will be on the examination, as this may make you even more nervous. Relax so that you can do your best!

What supplies will you need? First, be sure you bring your current original Amateur Radio license, if you have one. Bring a photocopy of your license, too, as well as the original and a photocopy of any Certificates of Successful Completion of Examination (CSCE) that you plan to use for exam credit. Bring along several sharpened number 2 pencils and two pens (blue or black ink). Be sure to have a good eraser. A pocket calculator may also come in handy. You may use a programmable calculator if that is the kind you have, but take it into your exam "empty" (cleared of all programs and constants in memory). Don't program equations ahead of time, because you may be asked to demonstrate that there is nothing in the calculator memory. The examining team has the right to refuse a candidate the

NCVEC QUICK-FORM 605 APPLICATION FOR
AMATEUR OPERATOR/PRIMARY STATION LICENSE

SECTION 1 - TO BE COMPLETED BY APPLICANT

PRINT LAST NAME: SOMMA | SUFFIX (Jr., Sr.) | FIRST NAME: MARIA | INITIAL | STATION CALL SIGN (IF ANY): KB1KJC

MAILING ADDRESS (Number and Street or P.O. Box): 225 MAIN ST. | SOCIAL SECURITY NUMBER (SSN) or (FRN) FCC FEDERAL REGISTRATION NUMBER: 0009876543

CITY: NEWINGTON | STATE CODE: CT | ZIP CODE (5 or 9 Numbers): 06111 | E-MAIL ADDRESS (OPTIONAL)

DAYTIME TELEPHONE NUMBER (Include Area Code) OPTIONAL | FAX NUMBER (Include Area Code) OPTIONAL | ENTITY NAME (IF CLUB, MILITARY RECREATION, RACES)

Type of Applicant: ☐ Individual ☐ Amateur Club ☐ Military Recreation ☐ RACES (Modify Only) | CLUB, MILITARY RECREATION, OR RACES CALL SIGN

I HEREBY APPLY FOR (Make an X in the appropriate box(es)) | SIGNATURE OF RESPONSIBLE CLUB OFFICIAL (not trustee)

☐ EXAMINATION for a new license grant
☐ EXAMINATION for upgrade of my license class
☐ CHANGE my name on my license to my new name
Former Name: ______________ (Last name) (Suffix) (First name) (MI)
☐ CHANGE my mailing address to above address
☐ CHANGE my station call sign systematically
Applicant's Initials: ______________
☐ RENEWAL of my license grant.

Do you have another license application on file with the FCC which has not been acted upon? | PURPOSE OF OTHER APPLICATION | PENDING FILE NUMBER (FOR VEC USE ONLY)

I certify that:

- I waive any claim to the use of any particular frequency regardless of prior use by license or otherwise;
- All statements and attachments are true, complete and correct to the best of my knowledge and belief and are made in good faith;
- I am not a representative of a foreign government;
- I am not subject to a denial of Federal benefits pursuant to Section 5301of the Anti-Drug Abuse Act of 1988, 21 U.S.C. § 862;
- The construction of my station will NOT be an action which is likely to have a significant environmental effect (See 47 CFR Sections 1.1301-1.1319 and Section 97.13(a));
- I have read and WILL COMPLY with Section 97.13(c) of the Commission's Rules regarding RADIOFREQUENCY (RF) RADIATION SAFETY and the amateur service section of OST/OET Bulletin Number 65.

Signature of applicant (Do not print, type, or stamp. Must match applicant's name above.) (Clubs: 2 different individuals must sign)
X Maria Somma Date Signed: 02-11-2010

SECTION 2 - TO BE COMPLETED BY ALL ADMINISTERING VEs

Applicant is qualified for operator license class:

☐ NO NEW LICENSE OR UPGRADE WAS EARNED
☐ TECHNICIAN Element 2
☐ GENERAL Elements 2 and 3
☒ AMATEUR EXTRA Elements 2, 3 and 4

DATE OF EXAMINATION SESSION: 02-11-2010
EXAMINATION SESSION LOCATION: NEWINGTON CT
VEC ORGANIZATION: ARRL
VEC RECEIPT DATE

I CERTIFY THAT I HAVE COMPLIED WITH THE ADMINISTERING VE REQUIRMENTS IN PART 97 OF THE COMMISSION'S RULES AND WITH THE INSTRUCTIONS PROVIDED BY THE COORDINATING VEC AND THE FCC.

VEs NAME (Print First, MI, Last, Suffix)	VEs STATION CALL SIGN	VEs SIGNATURE (Must match name)	DATE SIGNED
1st: Penny Harts	N1NAG	Penny Harts	2-11-10
2nd: RoseAnn Lawrence	KB1DMW	RoseAnn Lawrence	2-11-10
3rd: Perry Green	WY1O	Perry Green	02/11/10

DO NOT SEND THIS FORM TO FCC – THIS IS NOT AN FCC FORM.
IF THIS FORM IS SENT TO FCC, FCC WILL RETURN IT TO YOU WITHOUT ACTION.

NCVEC FORM 605 - February 2007
FOR VE/VEC USE ONLY - Page 1

Figure 3—At the test session, the Volunteer Examiners will help you fill out an NCVEC Quick Form 605, which will be filed with the FCC.

use of any calculator that they feel may contain information for the test or could otherwise be used to cheat on the exam.

The Volunteer Examiner team is required to check two forms of identification before you enter the test room. This includes your original Amateur Radio license, if you have one — not a photocopy. A photo ID of some type is best for the second form of ID, but is not required by the FCC. Other acceptable forms of identification include a driver's license, a piece of mail addressed to you or a birth certificate.

The following description of the testing procedure applies to exams coordinated by the ARRL/VEC, although many other VECs use a similar procedure.

Written Test

The examiner will give each applicant a test booklet, an answer sheet and scratch paper. You'll be shown where to sign your name and after that, you're on your own. The first thing to do is read the instructions.

Next, check the examination to see that all pages and questions are there. If not, report this to the examiner immediately. When filling in your answer sheet make sure your answers are marked next to the numbers that correspond to each question.

Go through the entire exam, and answer the easy questions first. Next, go back to the beginning and try the harder questions. Leave the really tough questions for last.

If you don't know the answer to a question, make your best guess. There is no additional penalty for answering incorrectly. If you have to guess, do it intelligently: At first glance, you may find that you can eliminate one or more "distractors." Of the remaining responses, more than one may seem correct; only one is the best answer, however. To the applicant who is fully prepared, incorrect distractors to each question are obvious. Nothing beats preparation!

After you've finished, check the examination thoroughly. You may have read a question wrong or goofed in your arithmetic. Don't be overconfident. There's no rush, so take your time. Think and check your answer sheet. When you feel you've done your best, return the test booklet, answer sheet and scratch pad to the examiner.

The Volunteer Examiner team will grade the exam while you wait. The passing mark is 74%. (That means 26 out of 35 questions correct with up to 9 incorrect answers on the Element 2 exam.) You will receive a Certificate of Successful Completion of Examination (CSCE — see **Figure 4**) showing all exam elements that you pass at that exam session. That certificate is valid for 365 days. Use it as proof that you passed those exam elements so you won't have to take them over again next time you take a license exam.

Forms and Procedures

To renew or modify a license, you can file a copy of FCC Form 605. In addition, hams who have held a valid license that has expired within the past two years may apply for reinstatement with an FCC Form 605.

Licenses are normally good for 10 years. Your application for a license renewal must be submitted to the FCC no more than 90 days before the license

American Radio Relay League VEC
Certificate of Successful Completion of Examination

ARRL The national association for AMATEUR RADIO

Test Site (City/State): 02-11-2010 Test Date: Newington, CT

NOTE TO VE TEAM: COMPLETELY CROSS OUT ALL BOXES BELOW THAT DO NOT APPLY TO THIS CANDIDATE.

The applicant named herein has presented valid proof for the exam element credit indicated below.
Pre 3/21/87 Technicians Element 3 credit

CREDIT for ELEMENTS PASSED VALID FOR 365 DAYS
You have passed the written element(s) indicated at right. You will be given credit for the appropriate examination element(s), for up to 365 days from the date shown at the top of this certificate.

EXAM ELEMENTS EARNED
~~Passed written Element 2~~
~~Passed written Element 3~~
Passed written Element 4

LICENSE UPGRADE NOTICE
If you also hold a valid FCC-issued Amateur radio license grant, this Certificate validates temporary operation with the operating privileges of your new operator class (see Section 97.9[b] of the FCC's Rules) until you are granted the license for your new operator class, or for a period of 365 days from the test date stated above on this certificate, whichever comes first.

NEW LICENSE CLASS EARNED
~~TECHNICIAN~~
~~GENERAL~~
EXTRA
~~NONE~~

LICENSE STATUS INQUIRIES
You can find out if a new license or upgrade has been "granted" by the FCC. For on-line inquiries see the FCC Web at **http://wireless.fcc.gov/uls/** ("Click on Search Licenses" button), or see the ARRL Web at **http://www.arrl.org/fcc/fcclook.php3**; or by calling FCC toll free at 888-225-5322; or by calling the ARRL at 1-860-594-0300 during business hours. **Allow 15 days from the test date before calling.**

THIS CERTIFICATE IS NOT A LICENSE, PERMIT, OR ANY OTHER KIND OF OPERATING AUTHORITY IN AND OF ITSELF. THE ELEMENT CREDITS AND/OR OPERATING PRIVILEGES THAT MAY BE INDICATED IN THE LICENSE UPGRADE NOTICE ARE VALID FOR 365 DAYS FROM THE TEST DATE. THE HOLDER NAMED HEREON MUST ALSO HAVE BEEN GRANTED AN AMATEUR RADIO LICENSE ISSUED BY THE FCC TO OPERATE ON THE AIR.

Candidate's Signature Maria Somma
Candidate's Name MARIA SOMMA Call Sign KB1KJC (If none, write none)
Address 225 MAIN ST.
City NEWINGTON State CT ZIP 06111

VE #1 Signature Call Sign N1NAG
VE #2 Signature Call Sign KB1DMW
VE #3 Signature Call Sign WY1O

COPIES: WHITE–Candidate, BLUE–Candidate, YELLOW–VE Team, PINK–ARRL/VEC
MVE 1/2009

Figure 4 — The CSCE (Certificate of Successful Completion of Examination) is your test session receipt that serves as proof that you have completed one or more exam elements. It can be used at other test sessions for 365 days.

expires. (We recommend you submit the application for renewal between 90 and 60 days before your license expires.) If the FCC receives your renewal application before the license expires, you may continue to operate until your new license arrives, even if it is past the expiration date.

If you forget to apply before your license expires, you may still be able to renew your license without taking another exam. There is a two-year grace period, during which you may apply for renewal of your expired license. Use an FCC Form 605 to apply for reinstatement (and your old call sign). If you apply for reinstatement of your expired license under this two-year grace period, you may not operate your station until your new license is issued.

If you move or change addresses, you should use an FCC Form 605 to notify the FCC of the change. If your license is lost or destroyed, however, just write a letter to the FCC explaining why you are requesting a new copy of your license.

You can ask one of the Volunteer Examiner Coordinators' offices to file your renewal application electronically if you don't want to mail the form to the FCC. You must still mail the form to the VEC, however. The ARRL/VEC office will electronically file application forms. This service is free for any ARRL member.

Electronic Filing

You can also file your license renewal or address modification using the FCC's Universal Licensing System (ULS) website, **www.fcc.gov/uls**. To use ULS, you must have an FCC Registration Number, or FRN. Obtain your FRN by registering with the Commission Registration System, known as CORES.

Described as an agency-wide registration system for anyone filing applications with or making payments to the FCC, CORES will assign a unique 10-digit

FCC Registration Number (FRN) to all registrants. All Commission systems that handle financial, authorization of service, and enforcement activities will use the FRN. The FCC says use of the FRN will allow it to more rapidly verify fee payment. Amateurs mailing payments to the FCC — for example as part of a vanity call sign application — would include their FRN on FCC Form 159.

The online filing system and further information about CORES is available by visiting the FCC web home page, **www.fcc.gov**, and clicking on the Commission Registration System link. Follow the directions on the website. It is also possible to register on CORES using a paper Form 160.

When you register with CORES you must supply a Taxpayer Identification Number, or TIN. For individuals, this is usually a Social Security Number. Club stations that do not have an EIN register as exempt. Anyone can register on CORES and obtain an FRN. You don't need a license to be registered.

Once you have registered on CORES and obtained your FRN, you can proceed to renew or modify your license using the Universal Licensing System by clicking on the "Online Filing" button. Follow the directions provided on the web page to connect to the FCC's ULS database.

Paper Filing

If you decide to "do the paperwork" on real paper instead of online, you'll need to get a blank FCC Form 605. This is not difficult! You can get FCC Form 605 with detailed instructions by contacting the FCC in any of these ways:

- FCC Forms Distribution Center, tel 800-418-3676.
- FCC Forms "Fax on Demand" — tel 202-418-0177, ask for form number 000605
- FCC Forms On-Line — **www.fcc.gov/formpage.html** or **ftp.fcc.gov/pub/Forms/Form605**

The ARRL/VEC has created a package that includes the portions of FCC Form 605 that are needed for amateur applications, as well as a condensed set of instructions for completing the form. Write to: ARRL/VEC, Form 605, 225 Main Street, Newington, CT 06111-1494. (Please include a large business-sized stamped self-addressed envelope with your request.)

And Now, Let's Begin

The complete Technician question pool (Element 2) is printed in this book. Each chapter lists all the questions for a particular subelement (such as Operating Procedures — T2). A brief explanation about the correct answer is given after each question.

Table 3 shows the study guide or syllabus for the Element 2 exam as released by the Volunteer-Examiner Coordinators' Question Pool Committee in January 2014. The syllabus lists the topics to be covered by the Technician exam, and so forms the basic outline for the remainder of this book. Use the syllabus to guide your study.

The question numbers used in the question pool refer to this syllabus. Each question number begins with a syllabus point number (for example, T0C or T1A). The question numbers end with a two-digit number. For example, question T3B09 is the ninth question about the T3B syllabus topics.

The Question Pool Committee designed the syllabus and question pool so there are the same number of topics in each subelement as there are exam questions from that subelement. For example, three exam questions on the Technician exam must be from the "Operating Procedures" subelement, so there are three groups for that topic. These are numbered T2A, T2B, and T2C. While not a requirement of the FCC Rules, the Question Pool Committee recommends that one question be taken from each group to make the best possible license exams.

Good luck with your studies!

Table 3

Technician Class (Element 2) Syllabus

SUBELEMENT T1 — FCC Rules, descriptions and definitions for the Amateur Radio Service, operator and station license responsibilities
[6 Exam Questions — 6 Groups]

T1A Amateur Radio Service: purpose and permissible use of the Amateur Radio Service; operator/primary station license grant; where FCC rules are codified; basis and purpose of FCC rules; meanings of basic terms used in FCC rules; interference; spectrum management

T1B Authorized frequencies: frequency allocations; ITU regions; emission modes; restricted sub-bands; spectrum sharing; transmissions near band edges

T1C Operator licensing: operator classes; sequential, special event, and vanity call sign systems; international communications; reciprocal operation; station license and licensee; places where the amateur service is regulated by the FCC; name and address on FCC license database; license term; renewal; grace period

T1D Authorized and prohibited transmission: communications with other countries; music; exchange of information with other services; indecent language; compensation for use of station; retransmission of other amateur signals; codes and ciphers; sale of equipment; unidentified transmissions; broadcasting

T1E Control operator and control types: control operator required; eligibility; designation of control operator; privileges and duties; control point; local, automatic and remote control; location of control operator

T1F Station identification; repeaters; third party communications; club stations; FCC inspection

SUBELEMENT T2 — Operating Procedures
[3 Exam Questions — 3 Groups]

T2A Station operation: choosing an operating frequency; calling another station; test transmissions; procedural signs; use of minimum power; choosing an operating frequency; band plans; calling frequencies; repeater offsets

T2B VHF/UHF operating practices: SSB phone; FM repeater; simplex; splits and shifts; CTCSS; DTMF; tone squelch; carrier squelch; phonetics; operational problem resolution; Q signals

T2C Public service: emergency and non-emergency operations; applicability of FCC rules; RACES and ARES; net and traffic procedures; emergency restrictions

SUBELEMENT T3 — Radio wave characteristics: properties of radio waves; propagation modes
[3 Exam Questions — 3 Groups]

T3A Radio wave characteristics: how a radio signal travels; fading; multipath; wavelength vs. penetration; antenna orientation

T3B Radio and electromagnetic wave properties: the electromagnetic spectrum; wavelength vs. frequency; velocity of electromagnetic waves; calculating wavelength

T3C Propagation modes: line of sight; sporadic E; meteor and auroral scatter and reflections; tropospheric ducting; F layer skip; radio horizon

SUBELEMENT T4 — Amateur radio practices and station set up
[2 Exam Questions — 2 Groups]

T4A Station setup: connecting microphones; reducing unwanted emissions; power source; connecting a computer; RF grounding; connecting digital equipment; connecting an SWR meter

T4B Operating controls: tuning; use of filters; squelch function; AGC; repeater offset; memory channels

SUBELEMENT T5 — Electrical principles: math for electronics; electronic principles; Ohm's Law
[4 Exam Questions — 4 Groups]

T5A Electrical principles, units, and terms: current and voltage; conductors and insulators; alternating and direct current

T5B Math for electronics: conversion of electrical units; decibels; the metric system

T5C Electronic principles: capacitance; inductance; current flow in circuits; alternating current; definition of RF; DC power calculations; impedance

T5D Ohm's Law: formulas and usage

SUBELEMENT T6 — Electrical components: semiconductors; circuit diagrams; component functions
[4 Exam Questions — 4 Groups]

T6A Electrical components: fixed and variable resistors; capacitors and inductors; fuses; switches; batteries

T6B Semiconductors: basic principles and applications of solid state devices; diodes and transistors

T6C Circuit diagrams; schematic symbols

T6D Component functions: rectification; switches; indicators; power supply components; resonant circuit; shielding; power transformers; integrated circuits

SUBELEMENT T7 — Station equipment: common transmitter and receiver problems; antenna measurements; troubleshooting; basic repair and testing
[4 Exam Questions — 4 Groups]

T7A Station equipment: receivers; transmitters; transceivers; modulation; transverters; low power and weak signal operation; transmit and receive amplifiers

T7B Common transmitter and receiver problems: symptoms of overload and overdrive; distortion; causes of interference; interference and consumer electronics; part 15 devices; over and under modulation; RF feedback; off frequency signals; fading and noise; problems with digital communications interfaces

T7C Antenna measurements and troubleshooting: measuring SWR; dummy loads; coaxial cables; feed line failure modes

T7D Basic repair and testing: soldering; using basic test instruments; connecting a voltmeter, ammeter, or ohmmeter

SUBELEMENT T8 — Modulation modes: amateur satellite operation; operating activities; non-voice communications
[4 Exam Questions — 4 Groups]

T8A Modulation modes: bandwidth of various signals; choice of emission type

T8B Amateur satellite operation; Doppler shift, basic orbits, operating protocols; control operator, transmitter power considerations; satellite tracking; digital modes

T8C Operating activities: radio direction finding; radio control; contests; linking over the Internet; grid locators

T8D Non-voice communications: image signals; digital modes; CW; packet; PSK31; APRS; error detection and correction; NTSC

SUBELEMENT T9 — Antennas and feed lines
[2 Exam Questions 2 Groups]

T9A Antennas: vertical and horizontal polarization; concept of gain; common portable and mobile antennas; relationships between antenna length and frequency

T9B Feed lines: types of feed lines; attenuation vs. frequency; SWR concepts; matching; weather protection; choosing RF connectors and feed lines

SUBELEMENT T0 — Electrical safety: AC and DC power circuits; antenna installation; RF hazards
[3 Exam Questions — 3 Groups]

T0A Power circuits and hazards: hazardous voltages; fuses and circuit breakers; grounding; lightning protection; battery safety; electrical code compliance

T0B Antenna safety: tower safety; erecting an antenna support; overhead power lines; installing an antenna

T0C RF hazards: radiation exposure; proximity to antennas; recognized safe power levels; exposure to others; radiation types; duty cycle

Subelement T1

FCC Rules

Your Technician exam (Element 2) will consists of 35 questions taken from the Technician question pool as prepared by the Volunteer Examiner Coordinator's Question Pool Committee. A certain number of questions are taken from each of the 10 subelements. There will be 6 questions from the FCC Rules subelement shown in this chapter. The questions are divided into 6 groups T1A through T1F.

The correct answer (A, B, C or D) is given in bold following the question and the possible responses at the beginning of an explanation section. This convention will be used throughout this book. In addition, at the end of each explanation you'll find the page number where this question is discussed in ARRL's *Ham Radio License Manual*, like this: [*Ham Radio License Manual*, page 7-3].

You'll often see a reference to Part 97 of the Federal Communications Commission rules set in brackets, like this: [97.3(a)(4)]. This tells you where to find the exact wording of the Rules as they relate to that question. You'll find the complete Part 97 Rules on the ARRL website at **www.arrl.org**.

SUBELEMENT T1 — FCC Rules, descriptions and definitions for the Amateur Radio Service, operator and station license responsibilities [6 Exam Questions — 6 Groups]

T1A Amateur Radio Service: purpose and permissible use of the Amateur Radio Service; operator/primary station license grant; where FCC rules are codified; basis and purpose of FCC rules; meanings of basic terms used in FCC rules; interference; spectrum management

T1A01 Which of the following is a purpose of the Amateur Radio Service as stated in the FCC rules and regulations?

A. Providing personal radio communications for as many citizens as possible
B. Providing communications for international non-profit organizations
C. Advancing skills in the technical and communication phases of the radio art
D. All of these choices are correct

C [97.1] — The FCC's Basis and Purpose for the Amateur service lists five reasons for creating the Amateur service, one of which is "advancing skills in the technical and communication phases of the radio art." [*Ham Radio License Manual*, page 7-2]

T1A02 **Which agency regulates and enforces the rules for the Amateur Radio Service in the United States?**

A. FEMA
B. The ITU
C. The FCC
D. Homeland Security

C [97.1] — Part 97 of the Federal Communication Commission's Rules governs the Amateur Radio Service in the United States. It is the FCC that enforces those rules. [*Ham Radio License Manual*, page 7-2]

T1A03 **Which part of the FCC regulations contains the rules governing the Amateur Radio Service?**

A. Part 73
B. Part 95
C. Part 90
D. Part 97

D The Amateur Service is defined by and operates according to the rules in Part 97 of the FCC's rules. [*Ham Radio License Manual*, page 7-1]

T1A04 **Which of the following meets the FCC definition of harmful interference?**

A. Radio transmissions that annoy users of a repeater
B. Unwanted radio transmissions that cause costly harm to radio station apparatus
C. That which seriously degrades, obstructs, or repeatedly interrupts a radio communication service operating in accordance with the Radio Regulations
D. Static from lightning storms

C [97.3(a)(23)] — A transmission that disturbs other authorized communications is called harmful interference. FCC Rules define harmful interference as, "Interference which endangers the functioning of a radionavigation service or of other safety communication service operating in accordance with the Radio Regulations." Not all interference is harmful. [*Ham Radio License Manual*, page 8-7]

T1A05 Which of the following is a purpose of the amateur service rules and regulations as defined by the FCC?

A. Enhancing international goodwill
B. Providing inexpensive communication for local emergency organizations
C. Training of operators in military radio operating procedures
D. All of these choices are correct

A [97.1(e)] — The amateur's ability to contact amateurs from other countries directly is a unique feature of Amateur Radio. [*Ham Radio License Manual*, page 7-2]

T1A06 Which of the following services are protected from interference by amateur signals under all circumstances?

A. Citizens Radio Service
B. Broadcast Service
C. Land Mobile Radio Service
D. Radionavigation Service

D [97.101(d), 97.303(o)(2)] — Because of the importance of radionavigation, amateurs must not interfere with those signals under any circumstances. [*Ham Radio License Manual*, page 8-7]

T1A07 What is the FCC Part 97 definition of telemetry?

A. An information bulletin issued by the FCC
B. A one-way transmission to initiate, modify or terminate functions of a device at a distance
C. A one-way transmission of measurements at a distance from the measuring instrument
D. An information bulletin from a VEC

C [97.3(a)(46)] — Related to telecommand signals (see question T1A13), telemetry signals are one-way transmissions as well, but send back measurements or status information from a measuring instrument or system. For example, a signal carrying the temperature of a repeater transmitter's enclosure is telemetry. [*Ham Radio License Manual*, page 6-33]

T1A08 Which of the following entities recommends transmit/receive channels and other parameters for auxiliary and repeater stations?

A. Frequency Spectrum Manager
B. Frequency Coordinator
C. FCC Regional Field Office
D. International Telecommunication Union

B [97.3(a)(22)] — Repeater input and output frequency pairs are fixed and have a common offset in each region. This enables the maximum number of repeaters to use the limited amount of spectrum. To keep order, a committee of volunteers known as a frequency coordinator recommends transmit and receive frequencies. [*Ham Radio License Manual*, page 7-16]

T1A09 Who selects a Frequency Coordinator?

A. The FCC Office of Spectrum Management and Coordination Policy
B. The local chapter of the Office of National Council of Independent Frequency Coordinators
C. Amateur operators in a local or regional area whose stations are eligible to be auxiliary or repeater stations
D. FCC Regional Field Office

C [97.3(a)(22)] — The frequency coordinator representatives are selected by the local or regional amateurs eligible to be auxiliary or repeater stations. This ensures participation by local amateurs and is a good example of self-regulation. [*Ham Radio License Manual*, page 7-17]

T1A10 What is the FCC Part 97 definition of an amateur station?

A. A station in the Amateur Radio Service consisting of the apparatus necessary for carrying on radio communications
B. A building where Amateur Radio receivers, transmitters, and RF power amplifiers are installed
C. Any radio station operated by a non-professional
D. Any radio station for hobby use

A [97.3(a)(5)] — The FCC defines an amateur station as, “A station licensed in the amateur service, including the apparatus necessary for carrying on radio communications.” A circular-sounding definition, but remember that the definitions are used in specific legal regulations and need to be precise. What the FCC is saying is that a station that conducts radio communications as required by the Amateur service rules in Part 97 meets the definition of an amateur station. [*Ham Radio License Manual*, page 7-3]

T1A11 When is willful interference to other amateur radio stations permitted?

A. Only if the station being interfered with is expressing extreme religious or political views
B. At no time
C. Only during a contest
D. At any time, amateurs are not protected from willful interference

B [97.101(d)] — Although harmful interference may happen accidentally, intentionally causing harmful interference to other communications is not allowed under any circumstances. [*Ham Radio License Manual*, page 8-8]

T1A12 Which of the following is a permissible use of the Amateur Radio Service?

A. Broadcasting music and videos to friends
B. Providing a way for amateur radio operators to earn additional income by using their stations to pass messages
C. Providing low-cost communications for start-up businesses
D. Allowing a person to conduct radio experiments and to communicate with other licensed hams around the world

D Communications in the Amateur Service must be free of commercial interests. [*Ham Radio License Manual*, page 7-2]

T1A13 What is the FCC Part 97 definition of telecommand?

A. An instruction bulletin issued by the FCC
B. A one-way radio transmission of measurements at a distance from the measuring instrument
C. A one-way transmission to initiate, modify or terminate functions of a device at a distance
D. An instruction from a VEC

C [97.3(a)(45)] — Telecommand signals are used to control models such as aircraft, boats, and vehicles, as well as space stations. [*Ham Radio License Manual*, page 6-33]

T1A14 **What must you do if you are operating on the 23 cm band and learn that you are interfering with a radiolocation station outside the United States?**

A. Stop operating or take steps to eliminate the harmful interference
B. Nothing, because this band is allocated exclusively to the amateur service
C. Establish contact with the radiolocation station and ask them to change frequency
D. Change to CW mode, because this would not likely cause interference

A [97.303(d)] — Not all US amateur allocations are allocated to amateurs worldwide. Where there are competing allocations and another service is considered the primary user, Amateur Radio is considered to be a secondary user. (See question T1B08.) [*Ham Radio License Manual*, page 7-16]

T1B Authorized frequencies: frequency allocations; ITU regions; emission modes; restricted sub-bands; spectrum sharing; transmissions near band edges

T1B01 **What is the ITU?**

A. An agency of the United States Department of Telecommunications Management
B. A United Nations agency for information and communication technology issues
C. An independent frequency coordination agency
D. A department of the FCC

B The International Telecommunication Union (ITU) is an international body of the United Nations that has responsibility for organizing the various radio services on a worldwide basis. This includes arranging international telecommunications treaties, as well as administrative responsibilities such as call signs and frequency allocations. [*Ham Radio License Manual*, page 7-17]

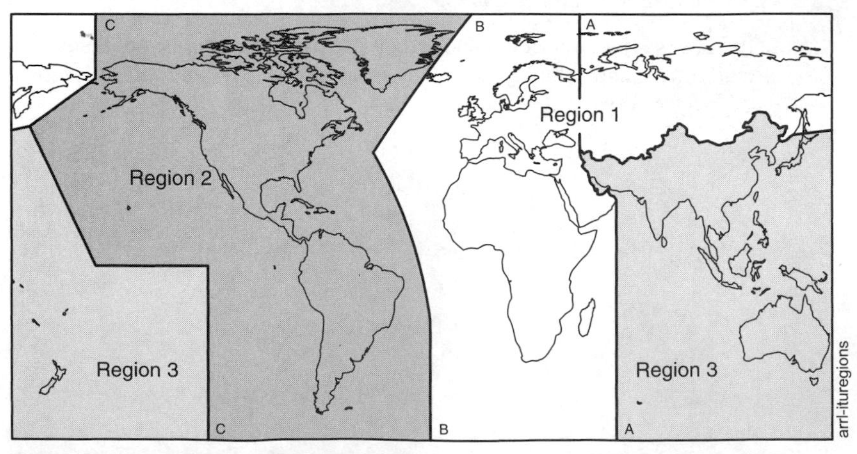

Figure T1-1 — This map shows the world divided into three International Telecommunication Union (ITU) Regions. The US is part of Region 2.

T1B02 **Why are the frequency assignments for some U.S. Territories different from those in the 50 U.S. States?**

A. Some U. S. Territories are located in ITU regions other than Region 2
B. Territorial governments are allowed to select their own frequency allocations
C. Territorial frequency allocations must also include those of adjacent countries
D. Any territory that was in existence before the ratification of the Communications Act of 1934 is exempt from FCC frequency regulations

A [97.301] — Frequency assignments may vary between the different ITU regions. [*Ham Radio License Manual*, page 7-18]

T1B03 **Which frequency is within the 6 meter band?**

A. 49.00 MHz
B. 52.525 MHz
C. 28.50 MHz
D. 222.15 MHz

B [97.301(a)] — The 6 meter band extends from 50.0 to 54.0 MHz in ITU Region 2, which includes North and South America. [*Ham Radio License Manual*, page 7-12]

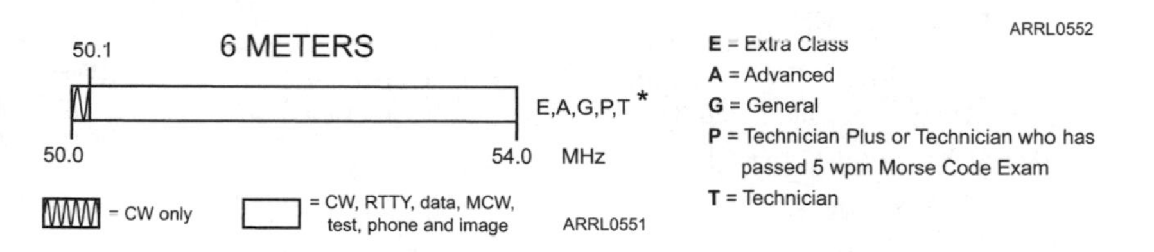

T1B04 Which amateur band are you using when your station is transmitting on 146.52 MHz?

A. 2 meter band
B. 20 meter band
C. 14 meter band
D. 6 meter band

A [97.301(a)] — The 2 meter band extends from 144.0 to 148.0 MHz in ITU Region 2, which includes North and South America. [*Ham Radio License Manual*, page 7-12]

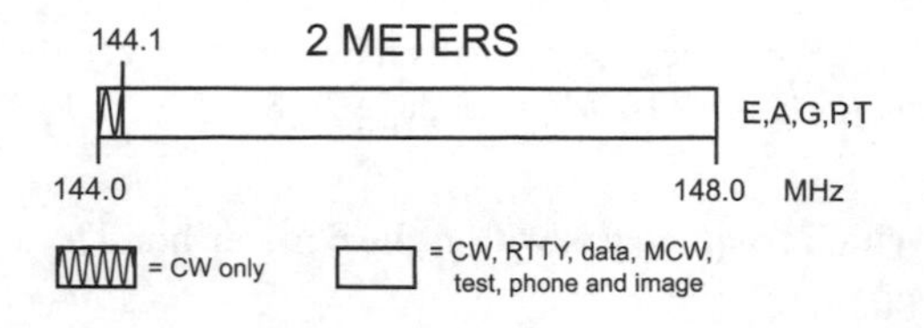

T1B05 Which 70 cm frequency is authorized to a Technician Class license holder operating in ITU Region 2?

A. 53.350 MHz
B. 146.520 MHz
C. 443.350 MHz
D. 222.520 MHz

C [97.301(a)] — The 70 centimeter band extends from 420.0 to 450.0 MHz in ITU Region 2, which includes North and South America. [*Ham Radio License Manual*, page 7-12]

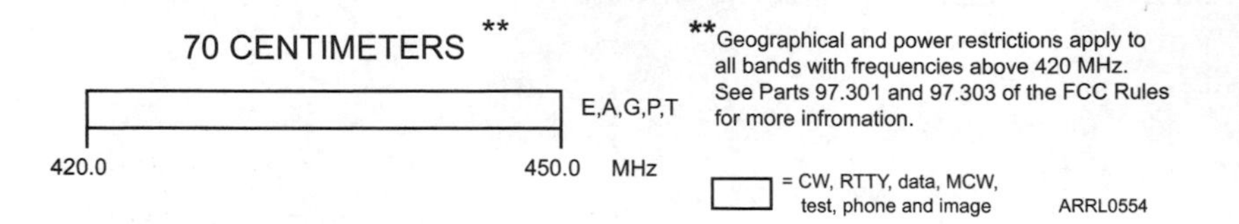

T1B06 **Which 23 cm frequency is authorized to a Technician Class licensee?**

A. 2315 MHz
B. 1296 MHz
C. 3390 MHz
D. 146.52 MHz

B [97.301(a)] — The 23 centimeter band extends from 1240 to 1300 MHz in ITU Region 2, which includes North and South America. [*Ham Radio License Manual*, page 7-12]

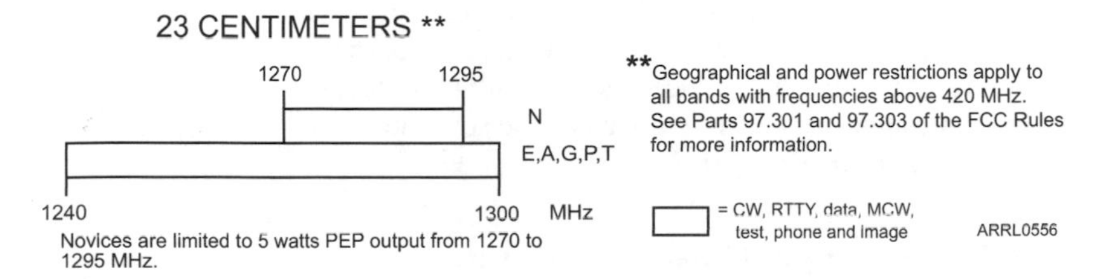

T1B07 **What amateur band are you using if you are transmitting on 223.50 MHz?**

A. 15 meter band
B. 10 meter band
C. 2 meter band
D. 1.25 meter band

D [97.301(a)] — The 1.25-meter band extends from 222 to 225 MHz in ITU Region 2, which includes North and South America. [*Ham Radio License Manual*, page 7-12]

***219-220 MHz allocated to amateurs on a secondary basis for fixed digital message forwarding systems only and can be operated by all licensees except Novices.

= CW, RTTY, data, MCW, test, phone and image ARRL0553

T1B08 Which of the following is a result of the fact that the amateur service is secondary in some portions of the 70 cm band?

A. U.S. amateurs may find non-amateur stations in the bands, and must avoid interfering with them
B. U.S. amateurs must give foreign amateur stations priority in those portions
C. International communications are not permitted on 70 cm
D. Digital transmissions are not permitted on 70 cm

A [97.303] — A radio service that is designated as the primary service on a band is protected from interference caused by other radio services. A radio service that is designated as the secondary service must not cause harmful interference to, and must accept interference from, stations in a primary service. The Amateur Service has many different frequency bands. Some of them are allocated on a primary basis and some are secondary. [*Ham Radio License Manual*, page 7-15]

T1B09 Why should you not set your transmit frequency to be exactly at the edge of an amateur band or sub-band?

A. To allow for calibration error in the transmitter frequency display
B. So that modulation sidebands do not extend beyond the band edge
C. To allow for transmitter frequency drift
D. All of these choices are correct

D [97.101(a), 97.301(a-e)] — Amateurs are allowed to use any frequency within a band, but you have to be careful when operating near the edge of the band. The rules require that all of your signal must remain within the band. Since, for example, on phone your radio displays the carrier frequency, you must leave room for the signal's sidebands. That means that if an FM voice signal is 15 kHz wide, your carrier frequency (in the center of the signal) should never be less than 7.5 kHz from the band edge. [*Ham Radio License Manual*, page 2-10]

T1B10 Which of the bands above 30 MHz that are available to Technician Class operators have mode-restricted sub-bands?

A. The 6 meter, 2 meter, and 70 cm bands
B. The 2 meter and 13 cm bands
C. The 6 meter, 2 meter, and 1.25 meter bands
D. The 2 meter and 70 cm bands

C [97.301(e), 97.305(c)] — Above 50 MHz, the only bands that are divided in any way are 6 meters, 2 meters and 1.25 meters. Both 6 meters and 2 meters have small CW-only sub-bands. 1.25 meters has a separate sub-band from 219-220 MHz in which data signals for fixed message forwarding systems are permitted. Otherwise, there are no mode restrictions in and above the 70 cm band. All of the separation of modes on those bands is according to band plans. [*Ham Radio License Manual*, page 7-13]

T1B11 What emission modes are permitted in the mode-restricted sub-bands at 50.0 to 50.1 MHz and 144.0 to 144.1 MHz?

A. CW only
B. CW and RTTY
C. SSB only
D. CW and SSB

A [97.301(a), 97.305 (a)(c)] — A small CW segment is set aside for beacon stations and types of operation that involve very weak CW signals, such as moonbounce and meteor scatter. [*Ham Radio License Manual*, page 7-13]

T1B12 Why are frequency assignments for U.S. stations operating maritime mobile not the same everywhere in the world?

A. Amateur maritime mobile stations in international waters must conform to the frequency assignments of the country nearest to their vessel
B. Amateur frequency assignments can vary among the three ITU regions
C. Frequency assignments are determined by the captain of the vessel
D. Amateur frequency assignments are different in each of the 90 ITU zones

B [97.301] — Frequency assignments may vary between the different ITU regions. [*Ham Radio License Manual*, page 7-18]

T1B13 Which emission may be used between 219 and 220 MHz?

A. Spread spectrum
B. Data
C. SSB voice
D. Fast-scan television

B [97.305(c)] — Only digital message forwarding stations may operate in this segment of the 1.25 meter band. [*Ham Radio License Manual*, page 7-13]

T1C Operator licensing: operator classes; sequential, special event, and vanity call sign systems; international communications; reciprocal operation; station license and licensee; places where the amateur service is regulated by the FCC; name and address on FCC license database; license term; renewal; grace period

T1C01 **Which type of call sign has a single letter for both the prefix and suffix?**

A. Vanity
B. Sequential
C. Special event
D. In-memoriam

C [97.3(a)(11)(iii)] — Any FCC-licensed amateur or club can also obtain a special 1×1 (1-by-1) call sign such as W3X or K6P for a short-duration special event of significance to the amateur community. These call signs are special because they only have one letter in both the prefix and suffix. [*Ham Radio License Manual*, page 7-22]

T1C02 **Which of the following is a valid US amateur radio station call sign?**

A. KMA3505
B. W3ABC
C. KDKA
D. 11Q1176

B Every amateur call sign has a prefix and a suffix. In the US, an amateur call sign prefix consists of one or two letters and one numeral. The suffix consists of one to three letters. [*Ham Radio License Manual*, page 7-20]

T1C03 **What types of international communications are permitted by an FCC-licensed amateur station?**

A. Communications incidental to the purposes of the amateur service and remarks of a personal character
B. Communications incidental to conducting business or remarks of a personal nature
C. Only communications incidental to contest exchanges, all other communications are prohibited
D. Any communications that would be permitted on an international broadcast station

A [97.117] — International treaties limit amateur communications to keep commercial and amateur communications separate. [*Ham Radio License Manual*, page 7-19]

T1C04 When are you allowed to operate your amateur station in a foreign country?

A. When the foreign country authorizes it
B. When there is a mutual agreement allowing third party communications
C. When authorization permits amateur communications in a foreign language
D. When you are communicating with non-licensed individuals in another country

A [97.107] — Amateur Radio is not authorized by all countries! If Amateur Radio is authorized, you must be licensed according to that country's regulations. Your US amateur license may suffice if there is a reciprocal operating agreement between the US and the foreign country, enabling you to operate your radio according to the rules of that country. There are three types of agreement: the International Amateur Radio Permit (IARP), the European Conference of Postal and Telecommunications Administration (CEPT) agreement, and international reciprocal agreements between the foreign country and the US. [*Ham Radio License Manual*, page 7-18]

T1C05 Which of the following is a vanity call sign which a Technician class amateur operator might select if available?

A. K1XXX
B. KA1X
C. W1XX
D. All of these choices are correct

A Technician class licensees can select call signs from Group C (1x3) and Group D (2x3). [*Ham Radio License Manual*, page 7-22]

T1C06 From which of the following locations may an FCC-licensed amateur station transmit, in addition to places where the FCC regulates communications?

A. From within any country that belongs to the International Telecommunication Union
B. From within any country that is a member of the United Nations
C. From anywhere within in ITU Regions 2 and 3
D. From any vessel or craft located in international waters and documented or registered in the United States

D [97.5(a)(2)] — Vessels and crafts documented or registered in the United States are bound by US rules and regulations. In addition, you must have permission to transmit and when you are inside a country's national boundaries, including territorial waters, you are required to operate according to their rules. [*Ham Radio License Manual*, page 7-18]

T1C07 What may result when correspondence from the FCC is returned as undeliverable because the grantee failed to provide the correct mailing address?

A. Fine or imprisonment
B. Revocation of the station license or suspension of the operator license
C. Require the licensee to be re-examined
D. A reduction of one rank in operator class

B [97.23] — The FCC requires you to maintain a valid current mailing address in their database at all times. This is so you can be contacted by mail, if needed. If you move or even change PO boxes, be sure to update your information using the FCC ULS online system. If you do not maintain a current address and mail to you is returned to the FCC as undeliverable, your license can be suspended or revoked and removed from the database. [*Ham Radio License Manual*, page 7-9]

T1C08 What is the normal term for an FCC-issued primary station/operator license grant?

A. Five years
B. Life
C. Ten years
D. Twenty years

C [97.25] — The FCC issues all amateur licenses for a 10-year term. You may renew your license for another 10-year term before it expires. [*Ham Radio License Manual*, page 7-8]

T1C09 What is the grace period following the expiration of an amateur license within which the license may be renewed?

A. Two years
B. Three years
C. Five years
D. Ten years

A [97.21(b)] — If you do forget to renew your license, you have up to two years to apply for renewal without having to take the exams again. After the two-year grace period, you will have to retake the exams. Your license is not valid during this two-year grace period, however, and you may not operate an amateur station with an expired license. All that the grace period implies is that the FCC will renew the license if you apply during that period. [*Ham Radio License Manual*, page 7-8]

T1C10 **How soon after passing the examination for your first amateur radio license may you operate a transmitter on an amateur service frequency?**

A. Immediately
B. 30 days after the test date
C. As soon as your operator/station license grant appears in the FCC's license database
D. You must wait until you receive your license in the mail from the FCC

C [97.5(a)] — Your Amateur Radio license is valid as soon as the FCC grants the license and posts the information about your license in the FCC database. You don't have to wait for the actual printed license to arrive by mail before you begin to transmit! You can check the FCC database using the FCC website or by using an Internet call sign lookup service. The ARRL maintains a call sign lookup service on the ARRL website, **www.arrl.org**. [*Ham Radio License Manual*, page 7-6]

T1C11 **If your license has expired and is still within the allowable grace period, may you continue to operate a transmitter on amateur service frequencies?**

A. No, transmitting is not allowed until the FCC database shows that the license has been renewed
B. Yes, but only if you identify using the suffix GP
C. Yes, but only during authorized nets
D. Yes, for up to two years

A [97.21(b)] — You may not transmit while your amateur license is expired. You must renew your license before you may transmit. (See also question T1C09.) [*Ham Radio License Manual*, page 7-8]

T1C12 **Who may select a desired call sign under the vanity call sign rules?**

A. Only licensed amateurs with general or extra class licenses
B. Only licensed amateurs with an extra class license
C. Only an amateur licensee who has been licensed continuously for more than 10 years
D. Any licensed amateur

D [97.19] — All amateurs are allowed to request vanity call signs from the call sign groups that are authorized to their license class. [*Ham Radio License Manual*, page 7-22]

T1C13 **For which license classes are new licenses currently available from the FCC?**

A. Novice, Technician, General, Advanced
B. Technician, Technician Plus, General, Advanced
C. Novice, Technician Plus, General, Advanced
D. Technician, General, Amateur Extra

D [97.9(a), 97.17(a)] — Novice and Advanced licenses are still active and may be renewed but no new licenses in those classes are being issued. Only Technician, General, and Extra class licenses are issue at this time. [*Ham Radio License Manual*, page 7-3]

T1C14 **Who may select a vanity call sign for a club station?**

A. Any Extra Class member of the club
B. Any member of the club
C. Any officer of the club
D. Only the person named as trustee on the club station license grant

D [97.21(a) (1)] — The trustee is responsible for administration of a club station license. [*Ham Radio License Manual*, page 7-22]

T1D Authorized and prohibited transmission: communications with other countries; music; exchange of information with other services; indecent language; compensation for use of station; retransmission of other amateur signals; codes and ciphers; sale of equipment; unidentified transmissions; broadcasting

T1D01 **With which countries are FCC-licensed amateur stations prohibited from exchanging communications?**

A. Any country whose administration has notified the ITU that it objects to such communications
B. Any country whose administration has notified the ARRL that it objects to such communications
C. Any country engaged in hostilities with another country
D. Any country in violation of the War Powers Act of 1934

A [97.111(a)(1)] — You may converse with amateurs in foreign countries unless either amateur's government prohibits the communications. (There are times when a government will not allow its amateurs to talk with people in other countries.) [*Ham Radio License Manual*, page 7-19]

T1D02 On which of the following occasions may an FCC-licensed amateur station exchange messages with a U.S. military station?

A. During an Armed Forces Day Communications Test
B. During a Memorial Day Celebration
C. During a Independence Day celebration
D. During a propagation test

A [97.111(a)(5)] — In general, hams can't communicate with non-amateur services, but the FCC may allow hams to talk to non-ham services at certain times or during a declared communications emergency. RACES operators may communicate with government stations during emergencies. The FCC also permits ham-to-military communication on Armed Forces Day. [*Ham Radio License Manual*, page 8-13]

T1D03 When is the transmission of codes or ciphers allowed to hide the meaning of a message transmitted by an amateur station?

A. Only during contests
B. Only when operating mobile
C. Only when transmitting control commands to space stations or radio control craft
D. Only when frequencies above 1280 MHz are used

C [97.211(b), 97.215(b)] — You can't use codes or ciphers (also known as encryption) to obscure the meaning of transmissions. This means you can't make up a "secret" code to send messages over the air to a friend. However, there are special exceptions. Control signals transmitted for remote control of model craft are not considered codes or ciphers. Neither are telemetry signals, such as a satellite might transmit to report on internal conditions. A space station (satellite) control operator may use specially coded signals to control the satellite. [*Ham Radio License Manual*, page 8-12]

T1D04 What is the only time an amateur station authorized to transmit music?

A. When incidental to an authorized retransmission of manned spacecraft communications
B. When the music produces no spurious emissions
C. When the purpose is to interfere with an illegal transmission
D. When the music is transmitted above 1280 MHz

A [97.113(a)(4), 97.113(c)] — Under FCC Rules, amateurs may not transmit music of any form. Retransmitting music from a broadcast program on a radio in your car or shack is also prohibited — so turn the radio down when you're on the air! There is one exception to the "No Music" rule. If you obtain special permission from NASA to retransmit the audio from the International Space Station for other amateurs to listen, and during that retransmission NASA or the astronauts play some music over the air, that's OK. [*Ham Radio License Manual*, page 8-13]

T1D05 When may amateur radio operators use their stations to notify other amateurs of the availability of equipment for sale or trade?

A. When the equipment is normally used in an amateur station and such activity is not conducted on a regular basis
B. When the asking price is $100.00 or less
C. When the asking price is less than its appraised value
D. When the equipment is not the personal property of either the station licensee or the control operator or their close relatives

A [97.113(a)(3)ii] — You are permitted to let other amateurs know that you have Amateur Radio equipment for sale or trade. This is a common practice on local "swap-and-shop" nets. The equipment or materials must be normally used in ham radio! You are also prohibited from selling or trading over the air on a regular basis. [*Ham Radio License Manual*, page 8-12]

T1D06 What, if any, are the restrictions concerning transmission of language that may be considered indecent or obscene?

A. The FCC maintains a list of words that are not permitted to be used on amateur frequencies
B. Any such language is prohibited
C. The ITU maintains a list of words that are not permitted to be used on amateur frequencies
D. There is no such prohibition

B [97.113(a)(4)] — Amateurs may not use obscene or indecent language — it is prohibited by the FCC Rules. Amateur Radio transmissions are public and anyone of any age can hear them. Depending on conditions and frequency, they might be heard anywhere in the world. There is no list of "words you can't say on Amateur Radio", but bear in mind the public nature of our communications and avoid any questionable language. [*Ham Radio License Manual*, page 8-11]

T1D07 What types of amateur stations can automatically retransmit the signals of other amateur stations?

A. Auxiliary, beacon, or Earth stations
B. Auxiliary, repeater, or space stations
C. Beacon, repeater, or space stations
D. Earth repeater, or space stations

B [97.113(d)] — Retransmitting the signals of another station is also generally prohibited, except when you are relaying messages or digital data from another station. Some types of stations (repeaters, auxiliary and space stations) are allowed to automatically retransmit signals on different frequencies or channels. [*Ham Radio License Manual*, page 8-13]

T1D08 In which of the following circumstances may the control operator of an amateur station receive compensation for operating the station?

A. When engaging in communications on behalf of their employer
B. When the communication is incidental to classroom instruction at an educational institution
C. When re-broadcasting weather alerts during a RACES net
D. When notifying other amateur operators of the availability for sale or trade of apparatus

B [97.113(a)(3)(iii)] — An exception to the prohibition of being compensated to operate is that teachers may use ham radio as part of their classroom instruction. In that case, they can be a control operator of a ham station, but it must be incidental to their job and can't be the majority of their duties. [*Ham Radio License Manual*, page 8-12]

T1D09 Under which of the following circumstances are amateur stations authorized to transmit signals related to broadcasting, program production, or news gathering, assuming no other means is available?

A. Only where such communications directly relate to the immediate safety of human life or protection of property
B. Only when broadcasting communications to or from the space shuttle.
C. Only where noncommercial programming is gathered and supplied exclusively to the National Public Radio network
D. Only when using amateur repeaters linked to the Internet

A [97.113(5)(b)] — Hams are specifically prohibited from assisting and participating in news gathering by broadcasting organizations, but when the immediate safety of human life or property is involved and normal communications are unavailable, they may use whatever type of communications are necessary. [*Ham Radio License Manual*, page 8-13]

T1D10 What is the meaning of the term "broadcasting" in the FCC rules for the amateur services?

A. Two-way transmissions by amateur stations
B. Transmission of music
C. Transmission of messages directed only to amateur operators
D. Transmissions intended for reception by the general public

D [97.3(a)(10)] — For the purposes of radio regulation, broadcasting means the transmission of information intended for reception by the general public. These broadcast transmissions may either be direct or relayed. Rules for Amateur Radio, a two-way communications service, prohibit broadcasting by amateur stations. [*Ham Radio License Manual*, page 8-13]

T1D11 When may an amateur station transmit without identifying?

A. When the transmissions are of a brief nature to make station adjustments
B. When the transmissions are unmodulated
C. When the transmitted power level is below 1 watt
D. When transmitting signals to control a model craft

D [97.119(a),97.215(a)] — Instead of transmitting a call sign, the station call sign and licensee's name and address must be displayed on the transmitter use to control a model craft. [*Ham Radio License Manual*, page 8-3]

T1D12 Question Withdrawn

T1E Control operator and control types: control operator required; eligibility; designation of control operator; privileges and duties; control point; local, automatic and remote control; location of control operator

T1E01 When is an amateur station permitted to transmit without a control operator?

A. When using automatic control, such as in the case of a repeater
B. When the station licensee is away and another licensed amateur is using the station
C. When the transmitting station is an auxiliary station
D. Never

D [97.7] — All amateur transmissions must be made under the supervision of a control operator whose job it is to be sure FCC Rules are followed. This is true even if the station is operating under automatic control. [*Ham Radio License Manual*, page 8-1]

T1E02 Who may a station licensee designate to be the control operator of an amateur station?

A. Any U.S. citizen or registered alien
B. Any family member of the station licensee
C. Any person over the age of 18
D. Only a person for whom an amateur operator/primary station license grant appears in the FCC database or who is authorized for alien reciprocal operation

D [97.7(a, b)] — A control operator must be named in the FCC amateur license database or be an alien with reciprocal operating authorization. (An alien is a citizen of another country.) This is a simple requirement — the FCC has to know who you are, that you are licensed, and where you can be contacted. Any licensed amateur can be a control operator. [*Ham Radio License Manual*, page 8-2]

T1E03 Who must designate the station control operator?

A. The station licensee
B. The FCC
C. The frequency coordinator
D. The ITU

A [97.103(b)] — The control operator doesn't have to be the station licensee and doesn't even have to be physically present at the transmitter, but all amateur transmissions are the responsibility of a control operator. The station licensee is responsible for designating the control operator. [*Ham Radio License Manual*, page 8-1]

T1E04 What determines the transmitting privileges of an amateur station?

A. The frequency authorized by the frequency coordinator
B. The class of operator license held by the station licensee
C. The highest class of operator license held by anyone on the premises
D. The class of operator license held by the control operator

D [97.105(b)] — An amateur station may transmit in any way permitted by the privileges of the control operator's license class. [*Ham Radio License Manual*, page 8-2]

T1E05 What is an amateur station control point?

A. The location of the station's transmitting antenna
B. The location of the station transmitting apparatus
C. The location at which the control operator function is performed
D. The mailing address of the station licensee

C [97.3(a)(14)] — This sounds like a circular definition, but the rule allows for an amateur station to be controlled from a remote location or even under automated control by devices and procedures that ensure proper operation. [*Ham Radio License Manual*, page 8-2]

T1E06 Under what type of control do APRS network digipeaters operate?

A. Automatic
B. Remote
C. Local
D. Manual

A [97.109(d)] — A digipeater relays digital messages from other stations just as a voice repeater relays phone transmissions from other stations. Since the digipeater can be activated at any time, the digipeater's control operator is required to place the station under automatic control. (See question T1E08) [*Ham Radio License Manual*, page 8-11]

T1E07 When the control operator is not the station licensee, who is responsible for the proper operation of the station?

A. All licensed amateurs who are present at the operation
B. Only the station licensee
C. Only the control operator
D. The control operator and the station licensee are equally responsible

D [97.103(a)] — Regardless of license class, both the control operator and station owner are responsible for proper operation of the station. The control operator is responsible for any transmissions made by the station and the licensee is responsible for controlling access to the station. [*Ham Radio License Manual*, page 8-2]

T1E08 Which of the following is an example of automatic control?

A. Repeater operation
B. Controlling the station over the Internet
C. Using a computer or other device to automatically send CW
D. Using a computer or other device to automatically identify

A [97.3(a)(6), 97.205(d)] — Under automatic control, the station operates completely under the control of devices and procedures that ensure the station complies with FCC rules. A control operator is still required, but need not be at the control point when the station is transmitting. [*Ham Radio License Manual*, page 8-11]

T1E09 What type of control is being used when the control operator is at the control point?

A. Radio control
B. Unattended control
C. Automatic control
D. Local control

D [97.109(b)] — Because the operator is directly manipulating the transmitter, this is local control. [*Ham Radio License Manual*, page 8-10]

T1E10 Which of the following is an example of remote control as defined in Part 97?

A. Repeater operation
B. Operating a station over the Internet
C. Controlling a model aircraft, boat or car by amateur radio
D. All of these choices are correct

B [97.3(a)(39)] — Because the transmitter is manipulated by the control operator through a control link, this is remote control. [*Ham Radio License Manual*, page 8-10]

T1E11 Who does the FCC presume to be the control operator of an amateur station, unless documentation to the contrary is in the station records?

A. The station custodian
B. The third party participant
C. The person operating the station equipment
D. The station licensee

D [97.103(a)] — The FCC will presume the station licensee to be the control operator unless there is a written record to the contrary. This is another reason why you should keep a logbook. [*Ham Radio License Manual*, page 8-2]

T1E12 When, under normal circumstances, may a Technician Class licensee be the control operator of a station operating in an exclusive Extra Class operator segment of the amateur bands?

A. At no time
B. When operating a special event station
C. As part of a multi-operator contest team
D. When using a club station whose trustee is an Extra Class operator licensee

A [97.119(e)] — If a Technician class licensee wishes to operate in an Extra class segment, an Extra class control operator must be present. [*Ham Radio License Manual*, page 8-2]

T1F Station identification; repeaters; third party communications; club stations; FCC inspection

T1F01 What type of identification is being used when identifying a station on the air as Race Headquarters?

A. Tactical call
B. An official call sign reserved for RACES drills
C. SSID
D. Broadcast station

A Tactical call signs (or tactical IDs) are used to help identify where a station is and what it is doing. Examples of tactical calls include "Waypoint 5," "First Aid Station," "Hollywood and Vine" and "Fire Watch on Coldwater Ridge." [*Ham Radio License Manual*, page 8-4]

T1F02 When using tactical identifiers such as "Race Headquarters" during a community service net operation, how often must your station transmit the station's FCC-assigned call sign?

A. Never, the tactical call is sufficient
B. Once during every hour
C. At the end of each communication and every ten minutes during a communication
D. At the end of every transmission

C [97.119(a)] — In addition to using the tactical identifier, identification using the station's FCC-assigned call sign is still required as stated in the FCC Rules. (See question T1F03.) [*Ham Radio License Manual*, page 8-4]

T1F03 When is an amateur station required to transmit its assigned call sign?

A. At the beginning of each contact, and every 10 minutes thereafter
B. At least once during each transmission
C. At least every 15 minutes during and at the end of a communication
D. At least every 10 minutes during and at the end of a communication

D [97.119(a)] — FCC regulations are very specific about station identification. You must identify your station every 10 minutes (or more frequently) during a contact and at the end of the communication. A communication may consist of several transmissions between stations and is not necessarily a single transmission. [*Ham Radio License Manual*, page 8-3]

T1F04 **Which of the following is an acceptable language to use for station identification when operating in a phone sub-band?**

A. Any language recognized by the United Nations
B. Any language recognized by the ITU
C. The English language
D. English, French, or Spanish

C [97.119(b)(2)] — You can use any language you want to communicate with other amateurs, even Esperanto! Amateur Radio provides you a great opportunity to practice your foreign language skills with other hams that speak that language. When you give your identification as a US station, however, you must use English. Foreign stations may have a similar requirement to use their native language when identifying their transmissions. [*Ham Radio License Manual*, page 8-4]

T1F05 **What method of call sign identification is required for a station transmitting phone signals?**

A. Send the call sign followed by the indicator RPT
B. Send the call sign using CW or phone emission
C. Send the call sign followed by the indicator R
D. Send the call sign using only phone emission

B [97.119(b)(2)] — The call sign must be transmitted using an emission type authorized for the transmitting frequency, so a station transmitting phone signals must identify its transmissions by means of spoken call signs or Morse code. The sending speed of a CW ID must not exceed 20 WPM (words per minute) if keyed by an automatic device used only for identification, a common practice for repeater stations. [*Ham Radio License Manual*, page 8-4]

T1F06 **Which of the following formats of a self-assigned indicator is acceptable when identifying using a phone transmission?**

A. KL7CC stroke W3
B. KL7CC slant W3
C. KL7CC slash W3
D. All of these choices are correct

D [97.119(c)] — An indicator is words, letters or numerals appended to and separated from your call sign during identification. When operating away from your home station, you should add an indicator to your call sign so that other stations are aware of your location. For instance, you might add the word "mobile" to your call sign when you are operating from your car on your way to work. This is important if using your regular call sign would be confusing. If Alaska resident KL7CC is operating from a location in the 3rd district, he would give his call sign as "KL7CC/W3." The added "/W3" is a self-assigned indicator. [*Ham Radio License Manual*, page 8-4]

T1F07 Which of the following restrictions apply when a non-licensed person is allowed to speak to a foreign station using a station under the control of a Technician Class control operator?

A. The person must be a U.S. citizen
B. The foreign station must be one with which the U.S. has a third party agreement
C. The licensed control operator must do the station identification
D. All of these choices are correct

B [97.115(a)(2)] — Allowing an unlicensed person to speak to a foreign station is called third-party traffic. Because the use of Amateur Radio by third-parties bypasses the usual communications networks, third-party traffic is only allowed between countries that specifically allow it. If no such agreement is in place, you may not exchange third-party traffic with that country. [*Ham Radio License Manual*, page 8-10]

T1F08 Which indicator is required by the FCC to be transmitted after a station call sign?

A. /M when operating mobile
B. /R when operating a repeater
C. / followed the FCC Region number when operating out of the region in which the license was issued
D. /KT, /AE or /AG when using new license privileges earned by CSCE while waiting for an upgrade to a previously issued license to appear in the FCC license database

D [97.119(f)] — You may use a self-assigned indicator at any time but the only FCC-required indicators are those that tell other stations what license class you have between the time you pass the exam and the time your new license class is published in the FCC database. [*Ham Radio License Manual*, page 8-5]

T1F09 What type of amateur station simultaneously retransmits the signal of another amateur station on a different channel or channels?

A. Beacon station
B. Earth station
C. Repeater station
D. Message forwarding station

C [97.3(a)(40)] — See question T1D07. [*Ham Radio License Manual*, page 2-12]

T1F10 Who is accountable should a repeater inadvertently retransmit communications that violate the FCC rules?

A. The control operator of the originating station
B. The control operator of the repeater
C. The owner of the repeater
D. Both the originating station and the repeater owner

A [97.205(g)] — Although there must be a control operator responsible for proper operation of the repeater, the primary responsibility for complying with FCC Rules for operating lies with the transmitting station. If repeater users consistently violate operating rules, the FCC can require that a repeater system be placed on remote control, meaning a control operator must monitor repeater operation. [*Ham Radio License Manual*, page 8-11]

T1F11 To which foreign stations do the FCC rules authorize the transmission of non-emergency third party communications?

A. Any station whose government permits such communications
B. Those in ITU Region 2 only
C. Those in ITU Regions 2 and 3 only
D. Those in ITU Region 3 only

A [97.115(a)] — There must be a third-party agreement in place between both governments for third-party communications to be permitted. [*Ham Radio License Manual*, page 8-9]

T1F12 How many persons are required to be members of a club for a club station license to be issued by the FCC?

A. At least 5
B. At least 4
C. A trustee and 2 officers
D. At least 2

B [97.5(b)(2)] — In order to ensure that a club actually exists, the FCC requires that it have at least four members, a name, be formally organized and managed and have a primary purpose devoted to Amateur Radio. [*Ham Radio License Manual*, page 7-22]

T1F13 **When must the station licensee make the station and its records available for FCC inspection?**

A. At any time ten days after notification by the FCC of such an inspection
B. At any time upon request by an FCC representative
C. Only after failing to comply with an FCC notice of violation
D. Only when presented with a valid warrant by an FCC official or government agent

B [97.103(c)] — As a station licensee, your responsibilities include the requirement to make the station and the station records available for inspection upon request by an FCC representative. An inspection can occur at any time, although they are actually quite rare. [*Ham Radio License Manual*, page 7-9]

Operating Procedures

SUBELEMENT T2 — Operating Procedures [3 Exam Questions — 3 Groups]

T2A Station operation: choosing an operating frequency; calling another station; test transmissions; procedural signs; use of minimum power; choosing an operating frequency; band plans; calling frequencies; repeater offsets

T2A01 What is the most common repeater frequency offset in the 2 meter band?

A. Plus 500 kHz
B. Plus or minus 600 kHz
C. Minus 500 kHz
D. Only plus 600 kHz

B The amount of repeater offset or shift is the same for almost all repeaters on one band as shown in the table below. This standardization makes it easy to use many different repeaters. Note that offset can often be either positive or negative by region or to allow more repeaters to occupy a single band, so be sure to use a repeater directory to determine the proper offset. [*Ham Radio License Manual*, page 6-16]

Standard Repeater Offsets by Band

Band	*Offset*
10 meters	–100 kHz
6 meters	Varies by region: –500 kHz, –1 MHz, –1.7 MHz
2 meters	+ or –600 kHz
1.25 meters	–1.6 MHz
70 cm	+ or –5 MHz
902 MHz	12 MHz
1296 MHz	12 MHz

T2A02 **What is the national calling frequency for FM simplex operations in the 70 cm band?**

A. 146.520 MHz
B. 145.000 MHz
C. 432.100 MHz
D. 446.000 MHz

D 446.000 MHz — the nationwide calling frequencies for FM simplex are listed in the table below. [*Ham Radio License Manual*, page 6-14]

FM Simplex Calling Frequencies

Band	*Frequency (MHz)*
10 meters	29.600
6 meters	52.525
2 meters	146.52
1.25 meters	223.50
70 cm	446.00
33 cm	906.50
23 cm	1294.50

T2A03 **What is a common repeater frequency offset in the 70 cm band?**

A. Plus or minus 5 MHz
B. Plus or minus 600 kHz
C. Minus 600 kHz
D. Plus 600 kHz

A See question T2A02. [*Ham Radio License Manual*, page 6-16]

T2A04 **What is an appropriate way to call another station on a repeater if you know the other station's call sign?**

A. Say break, break then say the station's call sign
B. Say the station's call sign then identify with your call sign
C. Say CQ three times then the other station's call sign
D. Wait for the station to call CQ then answer it

B To call another station when the repeater is not in use, just give both call signs. For example, "KA1JPA, this is N1OJS." If the repeater is in use, but the conversation sounds like it is about to end, wait before calling another station. [*Ham Radio License Manual*, page 6-12]

T2A05 How should you respond to a station calling CQ?

A. Transmit CQ followed by the other station's call sign
B. Transmit your call sign followed by the other station's call sign
C. Transmit the other station's call sign followed by your call sign
D. Transmit a signal report followed by your call sign

C On phone, say the other station's call sign followed by "this is" and your call sign at least once using phonetics, such as those in the table of ITU phonetics below. On CW, substitute the abbreviation DE for "this is." There is no need to send your call sign more than twice unless there is a lot of noise or interference present or the calling station is very weak. [*Ham Radio License Manual*, page 6-13]

ITU Phonetic Alphabet

Letter	*Word*	*Pronunciation*
A	Alfa	**AL** FAH
B	Bravo	**BRAH** VOH
C	Charlie	**CHAR** LEE
D	Delta	**DELL** TAH
E	Echo	**ECK** OH
F	Foxtrot	**FOKS** TROT
G	Golf	GOLF
H	Hotel	HOH **TELL**
I	India	**IN** DEE AH
J	Juliet	**JEW** LEE ETT
K	Kilo	**KEY** LOH
L	Lima	**LEE** MAH
M	Mike	MIKE
N	November	NO **VEM** BER
O	Oscar	**OSS** CAH
P	Papa	PAH **PAH**
Q	Quebec	KEH **BECK**
R	Romeo	**ROW** ME OH
S	Sierra	SEE **AIR** RAH
T	Tango	**TANG** GO
U	Uniform	**YOU** NEE FORM
V	Victor	**VIK** TAH
W	Whiskey	**WISS** KEY
X	X-Ray	**ECKS** RAY
Y	Yankee	**YANG** KEY
Z	Zulu	**ZOO** LOO

Note: The **boldfaced** syllables are emphasized. The pronunciations shown in this table were designed for those who speak any of the international languages. The pronunciations given for "Oscar" and "Victor" may seem awkward to English-speaking people in the US.

T2A06 What must an amateur operator do when making on-air transmissions to test equipment or antennas?

A. Properly identify the transmitting station
B. Make test transmissions only after 10:00 p.m. local time
C. Notify the FCC of the test transmission
D. State the purpose of the test during the test procedure

A Unidentified transmissions are not allowed, so be sure to follow FCC Rules. The identification rules are simple — give your call sign at least once every 10 minutes during a communication and when the communication is finished. If the transmission is too short for the 10 minute rule, just give your call sign as you end the transmission. [*Ham Radio License Manual*, page 8-6]

T2A07 Which of the following is true when making a test transmission into an antenna?

A. Station identification is not required if the transmission is less than 15 seconds
B. Station identification is not required if the transmission is less than 1 watt
C. Station identification is only required once an hour when the transmissions are for test purposes only
D. Station identification is required at least every ten minutes during the test and at the end

D See question T2A06. [*Ham Radio License Manual*, page 8-6]

T2A08 What is the meaning of the procedural signal "CQ"?

A. Call on the quarter hour
B. A new antenna is being tested (no station should answer]
C. Only the called station should transmit
D. Calling any station

D CQ literally means "Calling any station." You can usually tell good operators by how they call CQ. A good operator makes short, crisp calls separated by listening periods. Think of a CQ as an advertisement for your station and your operating skills. [*Ham Radio License Manual*, page 6-13]

T2A09 What brief statement is often used in place of "CQ" to indicate that you are listening on a repeater?

A. The words "Hello test" followed by your call sign
B. Your call sign
C. The repeater call sign followed by your call sign
D. The letters "QSY" followed by your call sign

B Since a repeater's signal is generally strong and the output frequency fixed, there is no need for an extended CQ call as there is when using SSB or CW. All that's necessary is to announce your presence and anyone who wishes to call you will then do so. It is also common to say "monitoring" following your call sign to reinforce that you are present and listening. [*Ham Radio License Manual*, page 6-11]

T2A10 What is a band plan, beyond the privileges established by the FCC?

A. A voluntary guideline for using different modes or activities within an amateur band
B. A mandated list of operating schedules
C. A list of scheduled net frequencies
D. A plan devised by a club to indicate frequency band usage

A Band plans are voluntary agreements between operators about how to use the bands under normal circumstances. These plans go into more detail on using different operating modes within an amateur band than is specified in FCC regulations. Good operators are familiar with the band plans and try to follow them. [*Ham Radio License Manual*, page 7-16]

T2A11 Which of the following is an FCC rule regarding power levels used in the amateur bands, under normal, non-distress circumstances?

A. There is no limit to power as long as there is no interference with other services
B. No more than 200 watts PEP may be used
C. Up to 1500 watts PEP may be used on any amateur frequency without restriction
D. While not exceeding the maximum power permitted on a given band, use the minimum power necessary to carry out the desired communication

D [97.313(a)] —The vast majority of contacts require far less power than the legal limit. Amateurs are required to avoid using excessive power levels to allow more hams to use the frequency. This does not mean you must reduce power until the other operator is barely able to hear you, just use a power level that provides satisfactory results. [*Ham Radio License Manual*, page 7-15]

T2A12 Which of the following is a guideline to use when choosing an operating frequency for calling CQ?

A. Listen first to be sure that no one else is using the frequency
B. Ask if the frequency is in use
C. Make sure you are in your assigned band
D. All of these choices are correct

D Listening is the amateur's most powerful tool for courteous and effective communication. A few seconds of listening often prevents an awkward intrusion on someone else's contact or a net operation. If you are pretty sure no stations are using the frequency, make a short "Is the frequency in use?" transmission. (If you are operating on a repeater, you won't be calling CQ, just proceed to call the station you want to contact.) And of course, you should be sure you will be operating in a band segment available to you! [*Ham Radio License Manual*, page 6-13]

T2B VHF/UHF operating practices: SSB phone; FM repeater; simplex; splits and shifts; CTCSS; DTMF; tone squelch; carrier squelch; phonetics; operational problem resolution; Q signals

T2B01 What is the term used to describe an amateur station that is transmitting and receiving on the same frequency?

A. Full duplex communication
B. Diplex communication
C. Simplex communication
D. Multiplex communication

C In Amateur Radio, simplex operation means that the stations are communicating by transmitting and receiving on the same frequency. Using a repeater with different transmit and receive frequencies is called *duplex* operation. [*Ham Radio License Manual*, page 6-9]

T2B02 What is the term used to describe the use of a sub-audible tone transmitted with normal voice audio to open the squelch of a receiver?

A. Carrier squelch
B. Tone burst
C. DTMF
D. CTCSS

D Repeater access tones were invented by Motorola to allow different commercial users to share a repeater without having to listen to each other's conversations. These tones are known by various names: Continuous Tone Coded Squelch System (CTCSS), PL (for Private Line, the Motorola trade name), or subaudible tones. FRS/GMRS radio users know these tones as privacy codes or privacy tones. [*Ham Radio License Manual*, page 6-16]

T2B03 Which of the following describes the muting of receiver audio controlled solely by the presence or absence of an RF signal?

A. Tone squelch
B. Carrier squelch
C. CTCSS
D. Modulated carrier

B To keep from having to listen to continuous noise when no signal is present, the squelch circuit was invented. The squelch circuit (sometimes called *carrier squelch*) mutes the receiver's audio output when no signal is present. [*Ham Radio License Manual*, page 5-7]

T2B04 Which of the following common problems might cause you to be able to hear but not access a repeater even when transmitting with the proper offset?

A. The repeater receiver may require an audio tone burst for access
B. The repeater receiver may require a CTCSS tone for access
C. The repeater receiver may require a DCS tone sequence for access
D. All of these choices are correct

D If you can hear a repeater's signal and you're sure you are using the right offset, but you can't access the repeater, then you probably don't have your radio set up to use the right type or frequency of access tone. [*Ham Radio License Manual*, page 6-17]

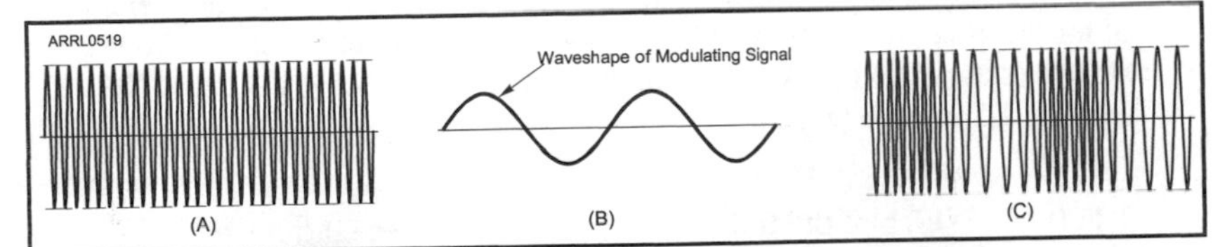

Figure T2-1 — At (A), each cycle of the unmodulated carrier has the same frequency. When the carrier is frequency-modulated with the signal at (B), its frequency increases and decreases corresponding to the increases and decreases in amplitude of the modulating signal.

T2B05 What determines the amount of deviation of an FM (as opposed to PM) signal?

A. Both the frequency and amplitude of the modulating signal
B. The frequency of the modulating signal
C. The amplitude of the modulating signal
D. The relative phase of the modulating signal

C The frequency of an FM signal varies only with the amplitude of the modulating signal as shown in Figure T2-1. The amount of variation is called *carrier deviation* or just deviation. [*Ham Radio License Manual*, page 2-10]

T2B06 What happens when the deviation of an FM transmitter is increased?

A. Its signal occupies more bandwidth
B. Its output power increases
C. Its output power and bandwidth increases
D. Asymmetric modulation occurs

A (See also question T2B05.) As deviation increases, so does the signal's bandwidth, so excessive deviation can cause interference to nearby signals. [*Ham Radio License Manual*, page 2-9]

T2B07 What could cause your FM signal to interfere with stations on nearby frequencies??

A. Microphone gain too high, causing over-deviation
B. SWR too high
C. Incorrect CTCSS Tone
D. All of these choices are correct

A Your equipment can cause spurious emissions if you operate it with some controls adjusted improperly. For example, if you set your microphone gain too high on an SSB transmitter, the resulting overmodulation creates spurious signals on nearby frequencies. On FM, misadjusting your tuning control or speaking too loudly (causing excessive deviation) can cause your signal to encroach on an adjacent channel. Every ham should make sure to transmit in a way that minimizes the possibility of causing harmful interference. Reports of interference such as transmitting off-frequency or generating spurious signals (splatter and buckshot) should always be checked out. [*Ham Radio License Manual*, page 2-9]

T2B08 Which of the following applies when two stations transmitting on the same frequency interfere with each other?

A. Common courtesy should prevail, but no one has absolute right to an amateur frequency
B. Whoever has the strongest signal has priority on the frequency
C. Whoever has been on the frequency the longest has priority on the frequency
D. The station which has the weakest signal has priority on the frequency

A Accidental interference happens frequently — perhaps you didn't hear the other stations before transmitting. Simple courtesy works wonders! Just say "excuse me," give your call sign, and move to a different frequency. [*Ham Radio License Manual*, page 8-7]

T2B09 Which of the following methods is encouraged by the FCC when identifying your station when using phone?

A. Use of a phonetic alphabet
B. Send your call sign in CW as well as voice
C. Repeat your call sign three times
D. Increase your signal to full power when identifying

A [97.119(b)(2)] — If the other operator is having difficulty copying your signals you should use the standard International Telecommunication Union (ITU) phonetic alphabet (see question T2A05). Use the words in the phonetic alphabet to spell out the letters in your call sign, your name or any other piece of information that might be confused if the letters are not received correctly. This phonetic alphabet is generally understood by hams in all countries. [*Ham Radio License Manual*, page 8-4]

T2B10 Which Q signal indicates that you are receiving interference from other stations?

A. QRM
B. QRN
C. QTH
D. QSB

A Q-signals are a system of making queries and exchanging information in an abbreviated form. They also allow operators who speak different languages to communicate. QRM refers to interference from other stations. QRN refers to interference from atmospheric static, and QTH means the station's location. QSB indicates signal fading. There are many useful Q-signals used by amateurs. [*Ham Radio License Manual*, page 6-5]

T2B11 Which Q signal indicates that you are changing frequency?

A. QRU
B. QSY
C. QSL
D. QRZ

B QSY means, "I am changing frequency." (See also question T2B10.) [*Ham Radio License Manual*, page 6-5]

T2B12 Under what circumstances should you consider communicating via simplex rather than a repeater?

A. When the stations can communicate directly without using a repeater
B. Only when you have an endorsement for simplex operation on your license
C. Only when third party traffic is not being passed
D. Only if you have simplex modulation capability

A Repeaters can be heard and accessed over a wide region, so if you and the other station can communicate via a direct simplex connection, it frees up the repeater for more people to use. [*Ham Radio License Manual*, page 6-14]

T2B13 Which of the following is true of the use of SSB phone in amateur bands above 50 MHz?

A. It is permitted only by holders of a General Class or higher license
B. It is permitted only on repeaters
C. It is permitted in at least some portion of all the amateur bands above 50 MHz
D. It is permitted only when power is limited to no more than 100 watts

C SSB is a very effective voice mode at low signal levels, such as when a repeater is not used. As a result, it is widely used for contacting distant (DX) stations beyond the line of sight range. [*Ham Radio License Manual*, page 6-9]

T2C Public service: emergency and non-emergency operations; applicability of FCC rules; RACES and ARES; net and traffic procedures; emergency restrictions

T2C01 When do the FCC rules NOT apply to the operation of an amateur station?

A. When operating a RACES station
B. When operating under special FEMA rules
C. When operating under special ARES rules
D. Never, FCC rules always apply

D [97.103(a)] — At all times, you are bound by FCC rules, even if using your radio in support of a public safety agency. If you are requested to use the station in a way that violates the FCC rules (except in the case of there being an immediate threat to life or property), you should politely decline. [*Ham Radio License Manual*, page 6-25]

T2C02 What is one way to recharge a 12-volt lead-acid station battery if the commercial power is out?

A. Cool the battery in ice for several hours
B. Add acid to the battery
C. Connect the battery in parallel with a vehicle's battery and run the engine
D. All of these choices are correct

C Your vehicle is an excellent source of backup power. It can easily recharge a storage battery. Smaller batteries should be carefully monitored for excessive temperature when charging from a vehicle since there is no circuitry to limit charging current. Follow standard vehicle safety practices when using your car to recharge batteries. [*Ham Radio License Manual*, page 5-18]

T2C03 What should be done to insure that voice message traffic containing proper names and unusual words are copied correctly by the receiving station?

A. The entire message should be repeated at least four times
B. Such messages must be limited to no more than 10 words
C. Such words and terms should be spelled out using a standard phonetic alphabet
D. All of these choices are correct

C Phonetics should be used whenever a word needs to be spelled out during a voice contact. (See also questions T2A05 and T2B09.) [*Ham Radio License Manual*, page 6-22]

T2C04 What do RACES and ARES have in common?

A. They represent the two largest ham clubs in the United States
B. Both organizations broadcast road and weather traffic information
C. Neither may handle emergency traffic supporting public service agencies
D. Both organizations may provide communications during emergencies

D RACES stands for Radio Amateur Civil Emergency Service, a communications service within the amateur service. RACES provides communications assistance to civil defense organizations in times of need. It is active only during periods of local, regional or national civil emergencies. ARES stands for Amateur Radio Emergency Service® and is sponsored by the ARRL. ARES presents a way for local amateurs to provide emergency communications while working with local public safety agencies and groups such as the Red Cross. ARES can provide communications assistance at any time. [*Ham Radio License Manual*, page 6-24]

T2C05 Which of the following describes the Radio Amateur Civil Emergency Service (RACES)?

A. A radio service using amateur frequencies for emergency management or civil defense communications
B. A radio service using amateur stations for emergency management or civil defense communications
C. An emergency service using amateur operators certified by a civil defense organization as being enrolled in that organization
D. All of these choices are correct

D [97.3(a)(38), 97.407] — See question T2C04. [*Ham Radio License Manual*, page 6-24]

T2C06 Which of the following is an accepted practice to get the immediate attention of a net control station when reporting an emergency?

A. Repeat the words SOS three times followed by the call sign of the reporting station
B. Press the push-to-talk button three times
C. Begin your transmission with "Priority" or "Emergency" followed by your call sign
D. Play a pre-recorded emergency alert tone followed by your call sign

C It is important to remember that no matter what the purpose or status of a net, a station with emergency traffic should break in at any time. If the net is operating on phone and you are reporting an emergency, break in by saying "Priority" or "Emergency," followed by your call sign. [*Ham Radio License Manual*, page 6-21]

T2C07 Which of the following is an accepted practice for an amateur operator who has checked into an emergency traffic net?

A. Provided that the frequency is quiet, announce the station call sign and location every 5 minutes
B. Move 5 kHz away from the net's frequency and use high power to ask other hams to keep clear of the net frequency
C. Remain on frequency without transmitting until asked to do so by the net control station
D. All of the choices are correct

C The desire to help may be strong, but remember that unnecessary transmissions just slow things down. The Net Control Station (NCS) will record your call sign and location so that if you're needed, you can be called. There's no need to remind the NCS that you're listening. Once you've checked in, you should not transmit unless you are specifically requested or authorized to do so or a request is made for capabilities or information that you can provide. [*Ham Radio License Manual*, page 6-21]

T2C08 Which of the following is a characteristic of good emergency traffic handling?

A. Passing messages exactly as received
B. Making decisions as to whether or not messages should be relayed or delivered
C. Communicating messages to the news media for broadcast outside the disaster area
D. All of these choices are correct

A The most important job during emergency and disaster net operation is the ability to accurately relay or "pass" messages exactly as they are received. [*Ham Radio License Manual*, page 6-22]

T2C09 Are amateur station control operators ever permitted to operate outside the frequency privileges of their license class?

A. No
B. Yes, but only when part of a FEMA emergency plan
C. Yes, but only when part of a RACES emergency plan
D. Yes, but only if necessary in situations involving the immediate safety of human life or protection of property

D [97.403] You are allowed to use any means of communication necessary if the immediate safety of human life or protection of property are at risk. [*Ham Radio License Manual*, page 6-25]

T2C10 What is the preamble in a formal traffic message?

A. The first paragraph of the message text
B. The message number
C. The priority handling indicator for the message
D. The information needed to track the message as it passes through the amateur radio traffic handling system

D Each message needs some information about the message and its content. The preamble gives the message a unique identity. The preamble consists of the following information:

Number — a unique number assigned by the station that creates the radiogram

Precedence — a description of the nature of the radiogram; Routine, Priority, Emergency and Welfare

Handling Instructions (HX) — for special instructions in how to the handle the radiogram.

Station of Origin — the call sign of the radio station from which the radiogram was first sent by Amateur Radio. (This allows information about the message to be returned to the sending station.)

Check — a count of the number of words or word equivalents in the text of the radiogram.

Place of Origin — the name of the town from which the radiogram started.

Time and Date — the time and date the radiogram is received at the station that first sent it.

Address — the complete name, street and number, city and state where the radiogram is going

[*Ham Radio License Manual*, page 6-22]

T2C11 What is meant by the term "check" in reference to a formal traffic message?

A. The check is a count of the number of words or word equivalents in the text portion of the message
B. The check is the value of a money order attached to the message
C. The check is a list of stations that have relayed the message
D. The check is a box on the message form that tells you the message was received

A See question T2C10. [*Ham Radio License Manual*, page 6-22]

T2C12 What is the Amateur Radio Emergency Service (ARES)?

A. Licensed amateurs who have voluntarily registered their qualifications and equipment for communications duty in the public service

B. Licensed amateurs who are members of the military and who voluntarily agreed to provide message handling services in the case of an emergency

C. A training program that provides licensing courses for those interested in obtaining an amateur license to use during emergencies

D. A training program that certifies amateur operators for membership in the Radio Amateur Civil Emergency Service

A (See also question T2C04) [Ham *Radio License Manual*, page 6-24]

Subelement T3

Radio Waves, Propagation and Antennas

SUBELEMENT T3 — Radio wave characteristics: properties of radio waves; propagation modes [3 Exam Questions — 3 Groups]

T3A Radio wave characteristics: how a radio signal travels; fading; multipath; wavelength vs. penetration; antenna orientation

T3A01 What should you do if another operator reports that your station's 2 meter signals were strong just a moment ago, but now they are weak or distorted?

A. Change the batteries in your radio to a different type
B. Turn on the CTCSS tone
C. Ask the other operator to adjust his squelch control
D. Try moving a few feet, as random reflections may be causing multi-path distortion

D In a location where reflections can occur, such as near vehicles or buildings, it is possible for a direct signal and a reflected signal to interfere with each other at the receiver and partially cancel. When this occurs, all that is necessary is to move approximately one-half wavelength in any direction — about three feet on the 2 meter band. It is likely that you'll find a "hot spot" within that range, or at least a location where the cancellation is greatly reduced. [*Ham Radio License Manual*, page 4-2]

T3A02 Why are UHF signals often more effective from inside buildings than VHF signals?

A. VHF signals lose power faster over distance
B. The shorter wavelength allows them to more easily penetrate the structure of buildings
C. This is incorrect; VHF works better than UHF inside buildings
D. UHF antennas are more efficient than VHF antennas

B Radio waves can penetrate openings in otherwise solid objects as long as at least one side of the opening is longer than about one-half wavelength. For this reason, the shorter wavelengths of UHF signals make them more effective at propagating into and out of buildings in urban areas. Obstructions also diffract or bend signals better as their wavelengths decrease, so UHF signals also have less shadowing effects in urban areas. [*Ham Radio License Manual*, page 4-2]

T3A03 What antenna polarization is normally used for long-distance weak-signal CW and SSB contacts using the VHF and UHF bands?

A. Right-hand circular
B. Left-hand circular
C. Horizontal
D. Vertical

C Horizontally polarized Yagis and quads are usually used for long-distance communications, especially for "weak signal" SSB and CW contacts on the VHF and UHF bands. Horizontal polarization is preferred because it results in lower ground losses when the wave reflects from or travels along the ground. [*Ham Radio License Manual*, page 4-15]

T3A04 What can happen if the antennas at opposite ends of a VHF or UHF line of sight radio link are not using the same polarization?

A. The modulation sidebands might become inverted
B. Signals could be significantly weaker
C. Signals have an echo effect on voices
D. Nothing significant will happen

B On the VHF and UHF bands, it is important to keep the transmitting and receiving antennas aligned so that they have matching polarizations. If the radio wave from a transmitter arrives at the receiver with a different polarization, the receiving antenna does not respond as well to the incoming radio wave. Since most repeaters have vertically polarized antennas, when using a low-power handheld radio to access a repeater it is important that your antenna be vertically polarized, too. [*Ham Radio License Manual*, page 4-6]

T3A05 When using a directional antenna, how might your station be able to access a distant repeater if buildings or obstructions are blocking the direct line of sight path?

A. Change from vertical to horizontal polarization
B. Try to find a path that reflects signals to the repeater
C. Try the long path
D. Increase the antenna SWR

B Remember that buildings and hills can act as radio reflectors. If you can "aim" your signal at one of these large reflectors, it is often possible for your signal to avoid an obstruction directly in your line of sight to the desired station. [*Ham Radio License Manual*, page 4-14]

T3A06 What term is commonly used to describe the rapid fluttering sound sometimes heard from mobile stations that are moving while transmitting?

A. Flip-flopping
B. Picket fencing
C. Frequency shifting
D. Pulsing

B Imagine yourself driving through an area with lots of reflections. You will travel in and out of "hot spots" and "dead spots" quite rapidly. The effect on your signal is an equally rapid increase and decrease in signal strength called mobile flutter or picket fencing. [*Ham Radio License Manual*, page 4-2]

T3A07 What type of wave carries radio signals between transmitting and receiving stations?

A. Electromagnetic
B. Electrostatic
C. Surface acoustic
D. Magnetostrictive

A An electromagnetic wave is a combination of an electric and a magnetic field, just like the fields in a capacitor and inductor but spreading out into space like ripples traveling across the surface of water. The wave's electric and magnetic fields oscillate at the same frequency as the RF current in the transmitting antenna. [*Ham Radio License Manual*, page 4-6]

T3A08 **Which of the following is a likely cause of irregular fading of signals received by ionospheric reflection?**

A. Frequency shift due to Faraday rotation
B. Interference from thunderstorms
C. Random combining of signals arriving via different paths
D. Intermodulation distortion

C Any time signals can take more than one path from the transmitter to the receiver, it is possible that they will interfere with each other and partially cancel, causing the combined signals to fade out. This can occur on any band, but it is common on the HF bands because of the different paths signals can take as they are refracted in the ionosphere and reflected from the Earth's surface. [*Ham Radio License Manual*, page 4-2]

T3A09 **Which of the following results from the fact that skip signals refracted from the ionosphere are elliptically polarized?**

A. Digital modes are unusable
B. Either vertically or horizontally polarized antennas may be used for transmission or reception
C. FM voice is unusable
D. Both the transmitting and receiving antennas must be of the same polarization

B As a radio wave travels through the ionosphere its polarization changes from vertical or horizontal to a combination of the two, called elliptical polarization. As a result, a receiving antenna of any polarization will respond to the incoming wave at least partially. This means both vertical and horizontal antennas are effective for receiving and transmitting on the HF bands where skip propagation is common. [*Ham Radio License Manual*, page 4-6]

T3A10 **What may occur if data signals propagate over multiple paths?**

A. Transmission rates can be increased by a factor equal to the number of separate paths observed
B. Transmission rates must be decreased by a factor equal to the number of separate paths observed
C. No significant changes will occur if the signals are transmitting using FM
D. Error rates are likely to increase

D Variations in signal strength from multipath that cause fading of voice signals can also cause digital data signals to be received with a higher error rate, particularly at VHF and UHF. (See questions T3A01 and T3A06.) [*Ham Radio License Manual*, page 4-2]

T3A11 Which part of the atmosphere enables the propagation of radio signals around the world?

A. The stratosphere
B. The troposphere
C. The ionosphere
D. The magnetosphere

C Radio waves at HF (and sometimes VHF) can be completely bent back towards the Earth by refraction in the ionosphere's F layers as if they were reflected. This is called sky wave propagation or skip. Since the Earth's surface is also conductive, it can also reflect radio waves. This means that a radio wave can be reflected between the ionosphere and ground multiple times. Each reflection from the ionosphere is called a hop and allows radio waves to be received hundreds or thousands of miles away. This is the most common way for hams to make long-distance contacts on the HF bands. [*Ham Radio License Manual*, page 4-3]

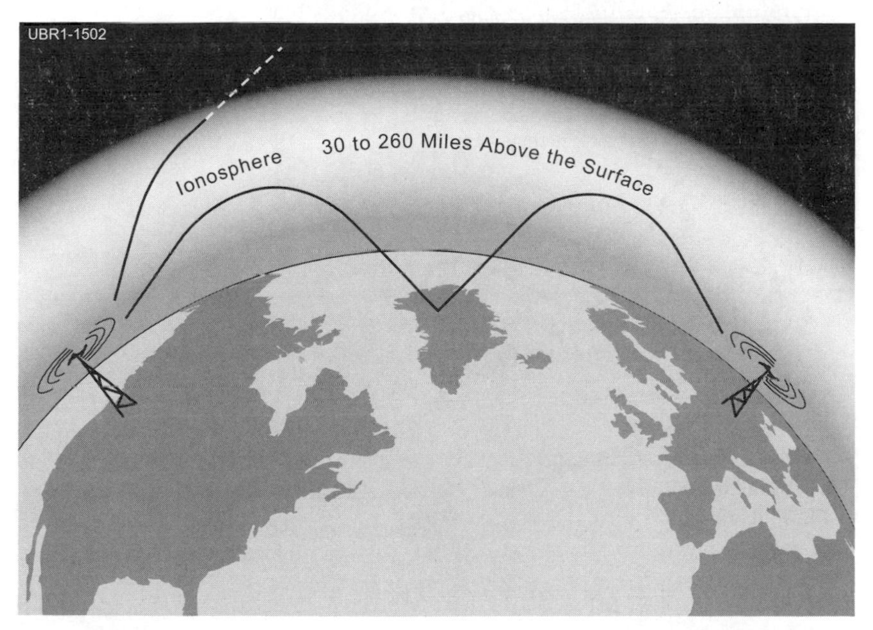

Figure T3-1 — The ionosphere is formed by solar ultraviolet (UV) radiation. The UV rays knock electrons loose from air molecules, creating weakly charged layers at different heights. These layers can absorb or refract radio signals, sometimes bending them back to the Earth.

T3B Radio and electromagnetic wave properties: the electromagnetic spectrum; wavelength vs. frequency; velocity of electromagnetic waves; calculating wavelength

T3B01 **What is the name for the distance a radio wave travels during one complete cycle?**

A. Wave speed
B. Waveform
C. Wavelength
D. Wave spread

C A radio wave travels at a constant speed, the speed of light. That means the radio wave will always cover the same distance during the time it takes the wave's oscillating electric and magnetic fields to make one complete cycle. That distance is the wave's wavelength. All radio waves of the same frequency, traveling at the speed of light, will have the same wavelength. [*Ham Radio License Manual*, page 2-5]

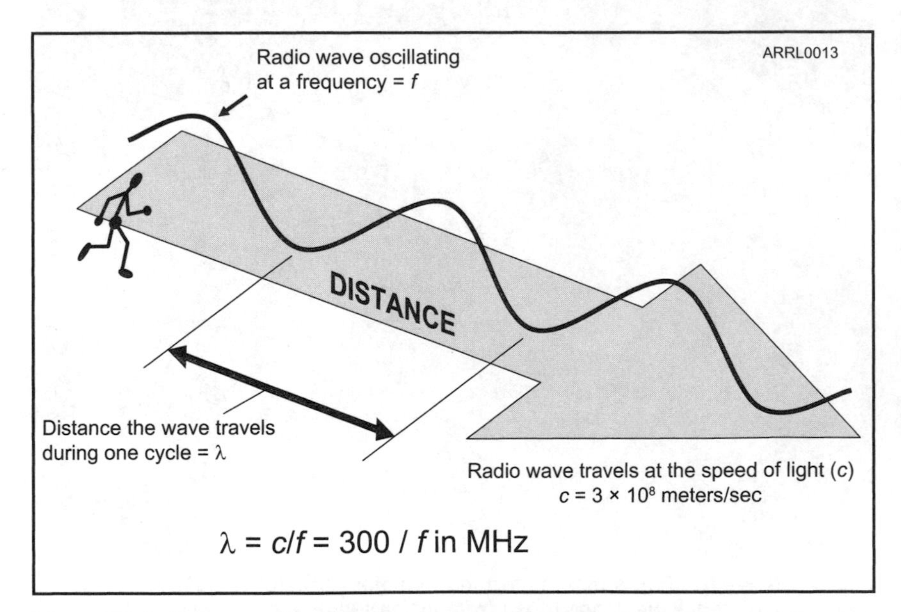

Figure T3-2 — As a radio wave travels, it oscillates at the frequency of the signal. The distance covered by the wave during the time it takes for one complete cycle is its wavelength, λ.

T3B02 What property of a radio wave is used to describe its polarization?

A. The orientation of the electric field
B. The orientation of the magnetic field
C. The ratio of the energy in the magnetic field to the energy in the electric field
D. The ratio of the velocity to the wavelength

A Because the radio wave's electric field is what causes electrons to move back and forth along an antenna to create current, the orientation of the electric field is the most important. (*Ham Radio License Manual*, page 4-6)

T3B03 What are the two components of a radio wave?

A. AC and DC
B. Voltage and current
C. Electric and magnetic fields
D. Ionizing and non-ionizing radiation

C See question T3A07. [*Ham Radio License Manual*, page 4-6]

T3B04 How fast does a radio wave travel through free space?

A. At the speed of light
B. At the speed of sound
C. Its speed is inversely proportional to its wavelength
D. Its speed increases as the frequency increases

A All electromagnetic energy — radio waves, light, X-rays — travels at the speed of light. In a vacuum, the speed of light (represented by a lower-case c) is 300,000,000 meters per second. It travels close to that speed in air. In denser materials such as water or glass, and in cables, light travels slower. [*Ham Radio License Manual*, page 2-5]

T3B05 How does the wavelength of a radio wave relate to its frequency?

A. The wavelength gets longer as the frequency increases
B. The wavelength gets shorter as the frequency increases
C. There is no relationship between wavelength and frequency
D. The wavelength depends on the bandwidth of the signal

B Because radio waves travel at a constant speed, the higher their frequency, the less time it takes to complete one cycle, and the less distance it has traveled during that cycle — its wavelength. Higher frequencies mean shorter wavelengths and vice versa. (See question T3A01.) [*Ham Radio License Manual*, page 2-5]

T3B06 What is the formula for converting frequency to approximate wavelength in meters?

A. Wavelength in meters equals frequency in hertz multiplied by 300
B. Wavelength in meters equals frequency in hertz divided by 300
C. Wavelength in meters equals frequency in megahertz divided by 300
D. Wavelength in meters equals 300 divided by frequency in megahertz

D The key to remembering this relationship is that wavelength and frequency are inversely proportional. That is, as one gets larger, the other must get smaller. The formula that describes this relationship is: Wavelength (λ, in meters) = 300 / (Frequency f, in MHz). [*Ham Radio License Manual*, page 2-6]

T3B07 What property of radio waves is often used to identify the different frequency bands?

A. The approximate wavelength
B. The magnetic intensity of waves
C. The time it takes for waves to travel one mile
D. The voltage standing wave ratio of waves

A Because of the radio wave's constant velocity (see question T3B01), knowing a radio wave's wavelength is the same as knowing its frequency and vice versa. It is traditional to refer to the amateur bands by an approximate wavelength. [*Ham Radio License Manual*, page 2-5]

T3B08 What are the frequency limits of the VHF spectrum?

A. 30 to 300 kHz
B. 30 to 300 MHz
C. 300 to 3000 kHz
D. 300 to 3000 MHz

B See the table of RF Spectrum Ranges. [*Ham Radio License Manual*, page 2-3]

RF Spectrum Ranges

Range Name	*Abbreviation*	*Frequency Range*
Very Low Frequency	VLF	3 kHz – 30 kHz
Low Frequency	LF	30 kHz – 300 kHz
Medium Frequency	MF	300 kHz – 3 MHz
High Frequency	HF	3 MHz – 30 MHz
Very High Frequency	VHF	30 MHz – 300 MHz
Ultra High Frequency	UHF	300 MHz – 3 GHz
Super High Frequency	SHF	3 GHz – 30 GHz
Extremely High Frequency	EHF	30 GHz – 300 GHz

T3B09 What are the frequency limits of the UHF spectrum?

A. 30 to 300 kHz
B. 30 to 300 MHz
C. 300 to 3000 kHz
D. 300 to 3000 MHz

D See the table of RF Spectrum Ranges. [*Ham Radio License Manual*, page 2-3]

T3B10 What frequency range is referred to as HF?

A. 300 to 3000 MHz
B. 30 to 300 MHz
C. 3 to 30 MHz
D. 300 to 3000 kHz

C See the table of RF Spectrum Ranges. [*Ham Radio License Manual*, page 2-3]

T3B11 What is the approximate velocity of a radio wave as it travels through free space?

A. 3000 kilometers per second
B. 300,000,000 meters per second
C. 300,000 miles per hour
D. 186,000 miles per hour

B See question T3B04. [*Ham Radio License Manual*, page 2-5]

T3C Propagation modes: line of sight; sporadic E; meteor and auroral scatter and reflections; tropospheric ducting; F layer skip; radio horizon

T3C01 Why are direct (not via a repeater) UHF signals rarely heard from stations outside your local coverage area?

A. They are too weak to go very far
B. FCC regulations prohibit them from going more than 50 miles
C. UHF signals are usually not reflected by the ionosphere
D. They collide with trees and shrubbery and fade out

C The ability of the ionosphere to refract or bend radio waves also depends on the frequency of the radio wave. Higher frequency waves are bent less than those of lower frequencies. At VHF and higher frequencies the waves usually pass through the ionosphere with only a little bending and are lost to space instead of returning to Earth where they can be received other stations. [*Ham Radio License Manual*, page 4-3]

T3C02 **Which of the following might be happening when VHF signals are being received from long distances?**

A. Signals are being reflected from outer space
B. Signals are arriving by sub-surface ducting
C. Signals are being reflected by lightning storms in your area
D. Signals are being refracted from a sporadic E layer

D At all points in the solar cycle, patches of the ionosphere's E layer can become sufficiently ionized to reflect VHF signals back to Earth. Most common on the 6 meter and 2 meter bands, sporadic E propagation occurs when signals are reflected from these patches, which form at irregular intervals and places, earning them the name "sporadic." A single sporadic E hop can be as long as 1500 miles. [*Ham Radio License Manual*, page 4-4]

T3C03 **What is a characteristic of VHF signals received via auroral reflection?**

A. Signals from distances of 10,000 or more miles are common
B. The signals exhibit rapid fluctuations of strength and often sound distorted
C. These types of signals occur only during winter nighttime hours
D. These types of signals are generally strongest when your antenna is aimed west

B The aurora (northern lights) is the glow from thin sheets of charged particles flowing down through the lower layers of the ionosphere at heights of 50 miles or more above the Earth's surface. Because the aurora is composed of charged particles, it reflects radio signals. The aurora is constantly changing, so the reflected signals change strength quickly and are often distorted. [*Ham Radio License Manual*, page 4-4]

T3C04 **Which of the following propagation types is most commonly associated with occasional strong over-the-horizon signals on the 10, 6, and 2 meter bands?**

A. Backscatter
B. Sporadic E
C. D layer absorption
D. Gray-line propagation

B See question T3C02. [*Ham Radio License Manual*, page 4-4]

T3C05 Which of the following effects might cause radio signals to be heard despite obstructions between the transmitting and receiving stations?

A. Knife-edge diffraction
B. Faraday rotation
C. Quantum tunneling
D. Doppler shift

A Radio waves can be reflected by any sudden change in the media through which they are traveling, such as a building, hill or even weather-related changes in the atmosphere. The figure shows how radio waves can also be diffracted or bent as they travel past sharp edges of these objects. This is called knife-edge diffraction. The resulting interference pattern creates radio shadows, especially at VHF and UHF frequencies. [*Ham Radio License Manual*, page 4-1]

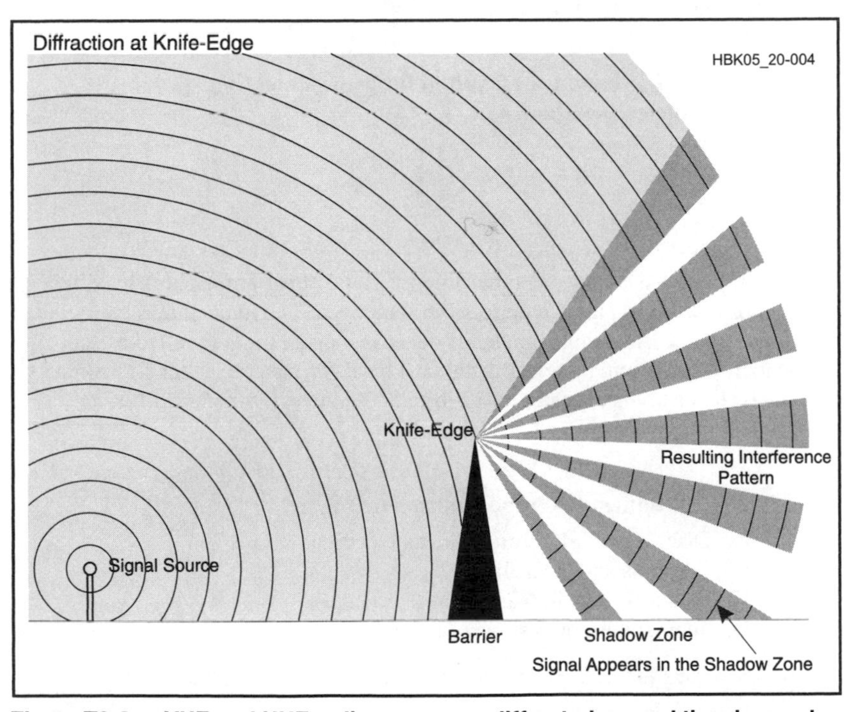

Figure T3-3 — VHF and UHF radio waves are diffracted around the sharp edge of a solid object, such as a building, hill or other obstruction. Some signals appear behind the obstruction as a result of interference between waves at the edge and those farther away. The resulting interference pattern creates a shadow zone of alternating strong and weak signal strength.

T3C06 What mode is responsible for allowing over-the-horizon VHF and UHF communications to ranges of approximately 300 miles on a regular basis?

A. Tropospheric scatter
B. D layer refraction
C. F2 layer refraction
D. Faraday rotation

A Propagation at and above VHF frequencies can be assisted by atmospheric phenomena such as weather fronts or temperature inversions. This is called tropospheric scatter propagation or just "tropo." (The troposphere is the lowest layer of the atmosphere.) Layers of air with different characteristics can also form structures called ducts that can guide even microwave signals for long distances. Tropo is regularly used by amateurs to make VHF and UHF contacts that would otherwise be impossible by line-of-sight propagation. Tropo contacts over 300-mile paths are not uncommon. [*Ham Radio License Manual*, page 4-2]

T3C07 What band is best suited for communicating via meteor scatter?

A. 10 meters
B. 6 meters
C. 2 meters
D. 70 cm

B A meteor is a meteoroid burning up in the upper atmosphere leaving a trail of ionized gas lasting up to several seconds. The ionized trail can reflect radio signals. Bouncing signals off of these ionized trails is called meteor scatter propagation. 6 meters is the best band for meteor scatter and contacts can be made at distances up to 1200 to 1500 miles. [*Ham Radio License Manual*, page 4-4]

T3C08 What causes tropospheric ducting?

A. Discharges of lightning during electrical storms
B. Sunspots and solar flares
C. Updrafts from hurricanes and tornadoes
D. Temperature inversions in the atmosphere

D See question T3C06. [*Ham Radio License Manual*, page 4-2]

T3C09 What is generally the best time for long-distance 10 meter band propagation via the F layer?

A. From dawn to shortly after sunset during periods of high sunspot activity

B. From shortly after sunset to dawn during periods of high sunspot activity

C. From dawn to shortly after sunset during periods of low sunspot activity

D. From shortly after sunset to dawn during periods of low sunspot activity

A The highest frequency signal that can be reflected back to a point on the Earth is the maximum usable frequency (MUF) between the transmitter and receiver. The MUF rises as the Sun illuminates the ionosphere. (The lowest frequency that can travel between those points without being absorbed is the lowest usable frequency (LUF) as shown in the figure.) That's why the upper HF bands, such as 10 meters, are more likely to be open during the day. In general, the higher sunspot activity is (creating higher UV levels to ionize the F layers) the higher the MUF will be between two points, often reaching the 10 meter band at 28-29.7 MHz. [*Ham Radio License Manual*, page 4-4]

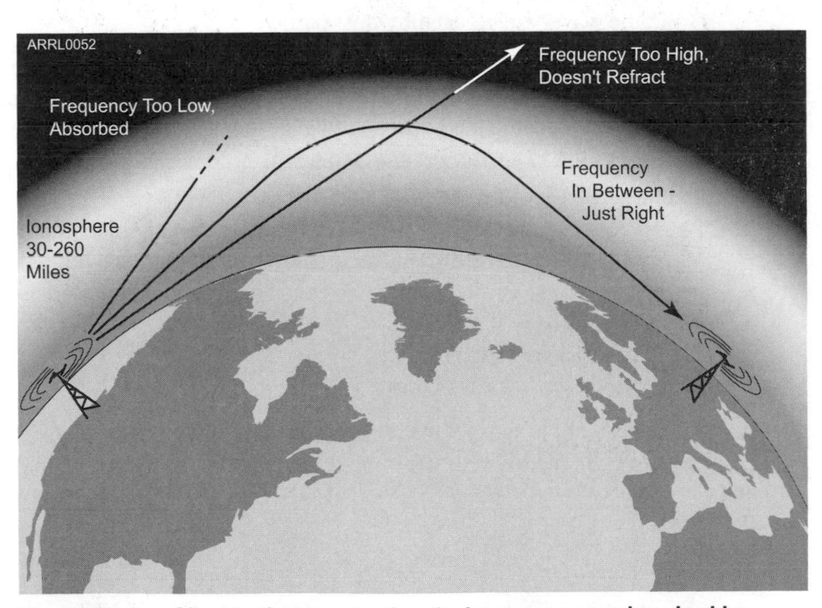

Figure T3-4 — Signals that are too low in frequency are absorbed by the ionosphere and lost. Signals that are too high in frequency pass through the ionosphere and are also lost. Signals in the right range of frequencies are refracted back toward the Earth and are received hundreds or thousands of miles away. Each F layer hop covers approximately 1500 to 2500 miles — the drawing here is not to scale.

T3C10 What is the radio horizon?

A. The distance over which two stations can communicate by direct path
B. The distance from the ground to a horizontally mounted antenna
C. The farthest point you can see when standing at the base of your antenna tower
D. The shortest distance between two points on the Earth's surface

A Since most propagation at VHF and UHF frequencies is line-of-sight, the limit of the range of these signals is called the radio horizon. (There is some slight bending of radio waves along the Earth's surface, so the radio horizon is somewhat more distant than the visual horizon.) Increasing the height of either the transmitting or receiving antennas also increases the radio horizon's distance. [*Ham Radio License Manual*, page 4-1]

T3C11 Why do VHF and UHF radio signals usually travel somewhat farther than the visual line of sight distance between two stations?

A. Radio signals move somewhat faster than the speed of light
B. Radio waves are not blocked by dust particles
C. The Earth seems less curved to radio waves than to light
D. Radio waves are blocked by dust particles

C Refraction bends VHF and UHF signals slightly back towards the Earth's surface. This allows the signals to travel farther than the visual line-of-sight horizon, in effect reducing the Earth's curvature to radio waves. (See question T3C10.) [*Ham Radio License Manual*, page 4-1]

T3C12 Which of the following bands may provide long distance communications during the peak of the sunspot cycle?

A. Six or ten meters
B. 23 centimeters
C. 70 centimeters or 1.25 meters
D. All of these choices are correct

A At the peak of the sunspot cycle, when sunspot activity and solar UV radiation are high, the MUF (see question T3C09) often reaches 10 meters and can even reach 6 meters on occasion. (*Ham Radio License Manual*, page 4-4)

Station Setup and Operation

SUBELEMENT T3 — Amateur radio practices and station set up [2 Exam Questions — 2 Groups]

T4A Station setup: connecting microphones; reducing unwanted emissions; power source; connecting a computer; RF grounding; connecting digital equipment; connecting an SWR meter

T4A01 Which of the following is true concerning the microphone connectors on amateur transceivers?

A. All transceivers use the same microphone connector type
B. Some connectors include push-to-talk and voltages for powering the microphone
C. All transceivers using the same connector type are wired identically
D. Un-keyed connectors allow any microphone to be connected

B The microphone connector of a transceiver is likely to include push-to-talk connections and also supply voltage for powering electret-style microphones. The wiring of the connector varies for the different manufacturers, so check the connections before trying a new microphone. [*Ham Radio License Manual*, page 5-6]

T4A02 How might a computer be used as part of an amateur radio station?

A. For logging contacts and contact information
B. For sending and/or receiving CW
C. For generating and decoding digital signals
D. All of these choices are correct

D Amateurs have incorporated the computer into nearly all aspects of ham radio. [*Ham Radio License Manual*, page 5-1]

T4A03 **Which is a good reason to use a regulated power supply for communications equipment?**

A. It prevents voltage fluctuations from reaching sensitive circuits
B. A regulated power supply has FCC approval
C. A fuse or circuit breaker regulates the power
D. Power consumption is independent of load

A To operate properly, communications equipment requires a power supply with a stable output voltage. Regulation prevents voltage fluctuations of the input power source from causing similar fluctuations at the output. [*Ham Radio License Manual*, page 5-15]

T4A04 **Where must a filter be installed to reduce harmonic emissions from your station?**

A. Between the transmitter and the antenna
B. Between the receiver and the transmitter
C. At the station power supply
D. At the microphone

A Every transmitter's RF output contains weak harmonics of the desired output signal and other spurious emissions that can cause interference to nearby equipment. In extreme cases, a misadjusted or defective transmitter can generate strong interfering signals. To prevent harmonics from being radiated, a low-pass or band-pass filter must be installed at the transmitter's connection to the antenna feed line. [*Ham Radio License Manual*, page 5-21]

T4A05 **Where should an in-line SWR meter be connected to monitor the standing wave ratio of the station antenna system?**

A. In series with the feed line, between the transmitter and antenna
B. In series with the station's ground
C. In parallel with the push-to-talk line and the antenna
D. In series with the power supply cable, as close as possible to the radio

A The SWR meter is inserted in the antenna system feed line (in series with the feed line). SWR can be measured anywhere in the feed line but the measurement is of most use at the transmitter output so that an antenna tuner can be adjusted for the lowest SWR. [*Ham Radio License Manual*, page 4-18]

T4A06 Which of the following would be connected between a transceiver and computer in a packet radio station?

A. Transmatch
B. Mixer
C. Terminal node controller
D. Antenna

C The terminal node controller (TNC) converts the audio signals from your radio's receiver output to data characters for the computer. Similarly, characters from the computer are converted to audio signals that your transmitter can use. The TNC may also act as the push-to-talk switch to turn the transmitter on and off. [*Ham Radio License Manual*, page 5-13]

T4A07 How is a computer's sound card used when conducting digital communications using a computer?

A. The sound card communicates between the computer CPU and the video display
B. The sound card records the audio frequency for video display
C. The sound card provides audio to the microphone input and converts received audio to digital form
D. All of these choices are correct

C Sound cards and their host computers have become sufficiently powerful that for several popular digital modes such as RTTY and PSK31, there is no need for a separate TNC or protocol controller. The sound card and software running on the host can perform the necessary conversions between audio signals and data characters. [*Ham Radio License Manual*, page 5-13]

T4A08 Which type of conductor is best to use for RF grounding?

A. Round stranded wire
B. Round copper-clad steel wire
C. Twisted-pair cable
D. Flat strap

D Connections to an RF ground bus should be made with a short, wide conductor such as copper flashing or strap or heavy solid wire (#8 AWG or larger). Solid strap is best because it presents the lowest impedance to RF current. [*Ham Radio License Manual*, page 5-25]

T4A09 Which of the following could you use to cure distorted audio caused by RF current flowing on the shield of a microphone cable?

A. Band-pass filter
B. Low-pass filter
C. Preamplifier
D. Ferrite choke

D RF choke or common-mode filters made of ferrite material are used to reduce RF currents flowing on unshielded wires such as speaker cables, ac power cords and telephone modular cords. Ferrite chokes are also used to reduce RF current flowing on the outside of shielded audio and computer cables. [*Ham Radio License Manual*, page 5-20]

T4A10 What is the source of a high-pitched whine that varies with engine speed in a mobile transceiver's receive audio?

A. The ignition system
B. The alternator
C. The electric fuel pump
D. Anti-lock braking system controllers

B Alternator whine is a type of noise caused by noise on the dc power system inside your own vehicle. You might hear it with the received audio but more likely it will be heard by others as a high-pitched whine on your transmitted signal that varies with your engine speed. It can be removed by a dc power filter at your radio. [*Ham Radio License Manual*, page 5-16]

T4A11 Where should the negative return connection of a mobile transceiver's power cable be connected?

A. At the battery or engine block ground strap
B. At the antenna mount
C. To any metal part of the vehicle
D. Through the transceiver's mounting bracket

A Connect the radio's negative lead to the negative battery terminal or where the battery ground lead is connected to the vehicle body. Do not rely on ground connections via the vehicle's body because vehicle sections may not be bonded together dependably. Many body components may be plastic or composite, as well. The most reliable negative power connection is directly to the battery or its ground strap. [*Ham Radio License Manual*, page 5-15]

T4A12 What could be happening if another operator reports a variable high-pitched whine on the audio from your mobile transmitter?

A. Your microphone is picking up noise from an open window
B. You have the volume on your receiver set too high
C. You need to adjust your squelch control
D. Noise on the vehicle's electrical system is being transmitted along with your speech audio

D See question T4A10. [*Ham Radio License Manual*, page 5-16]

T4B Operating controls: tuning; use of filters; squelch function; AGC; repeater offset; memory channels

T4B01 What may happen if a transmitter is operated with the microphone gain set too high?

A. The output power might be too high
B. The signal might become distorted
C. The frequency might vary
D. The SWR might increase

B Your transmitter will produce spurious emissions if your microphone gain is too high. For example, on an SSB transmitter the resulting overmodulation creates spurious signals on nearby frequencies. On FM, the result will be excessive deviation that can cause your signal to encroach on an adjacent channel. [*Ham Radio License Manual*, page 5-4]

T4B02 Which of the following can be used to enter the operating frequency on a modern transceiver?

A. The keypad or VFO knob
B. The CTCSS or DTMF encoder
C. The Automatic Frequency Control
D. All of these choices are correct

A The VFO (variable frequency oscillator) is the circuit that controls the frequency of operation on both receive and transmit. Many radios also have a numeric keypad that allows you to enter the desired frequency directly. The VFO is used when you are tuning across the band looking for a station or particular signal. Keypads are used when changing directly between known frequencies. [*Ham Radio License Manual*, page 5-2]

T4B03 What is the purpose of the squelch control on a transceiver?

A. To set the highest level of volume desired
B. To set the transmitter power level
C. To adjust the automatic gain control
D. To mute receiver output when no signal is being received

D In the absence of a signal on an FM receiver, you will hear noise. FM receivers have a squelch circuit that cuts off the speaker unless a signal is present, called "closing" the squelch. The squelch control adjusts the threshold at which squelch circuits turn the speaker on and off. The proper setting for the squelch control is just above the point at which the receiver audio is cut off. If set at a higher level, some signals will not be strong enough to "open" the squelch. [*Ham Radio License Manual*, page 5-7]

T4B04 What is a way to enable quick access to a favorite frequency on your transceiver?

A. Enable the CTCSS tones
B. Store the frequency in a memory channel
C. Disable the CTCSS tones
D. Use the scan mode to select the desired frequency

B By storing the frequency and any other pertinent information in a memory channel you can quickly return to the frequency at any time. [*Ham Radio License Manual*, page 5-2]

T4B05 Which of the following would reduce ignition interference to a receiver?

A. Change frequency slightly
B. Decrease the squelch setting
C. Turn on the noise blanker
D. Use the RIT control

C Ignition noise consists of a sharp pulse each time a spark plug in the engine fires. Just like the early spark radio transmitters, the arc produces RF signals across a wide range of frequencies. You hear the signals as buzzing or raspy noise that changes pitch along with engine speed. The noise blanker circuit in your radio detects these pulses and mutes the receiver for a short period. [*Ham Radio License Manual*, page 5-7]

T4B06 Which of the following controls could be used if the voice pitch of a single-sideband signal seems too high or low?

A. The AGC or limiter
B. The bandwidth selection
C. The tone squelch
D. The receiver RIT or clarifier

D Receiver incremental tuning (RIT) is a fine-tuning control used for SSB or CW operation. RIT allows the operator to adjust the receiver frequency without changing the transmitter frequency. This allows you to tune in a station that is slightly off frequency or to adjust the pitch of an operator's voice that seems too high or low. [*Ham Radio License Manual*, page 5-7]

T4B07 What does the term "RIT" mean?

A. Receiver Input Tone
B. Receiver Incremental Tuning
C. Rectifier Inverter Test
D. Remote Input Transmitter

B See question T4B06. [*Ham Radio License Manual*, page 5-7]

T4B08 What is the advantage of having multiple receive bandwidth choices on a multimode transceiver?

A. Permits monitoring several modes at once
B. Permits noise or interference reduction by selecting a bandwidth matching the mode
C. Increases the number of frequencies that can be stored in memory
D. Increases the amount of offset between receive and transmit frequencies

B IF filters are specified by bandwidth in Hz or kHz. Wide filters (around 2.4 kHz) are used for SSB reception on phone. Narrow filters (around 500 Hz) are used for Morse code and data mode reception. Having multiple filters available allows you to reduce noise or interference by selecting a filter with just enough bandwidth to pass the desired signal. [*Ham Radio License Manual*, page 5-7]

T4B09 **Which of the following is an appropriate receive filter bandwidth to select in order to minimize noise and interference for SSB reception?**

A. 500 Hz
B. 1000 Hz
C. 2400 Hz
D. 5000 Hz

C See question T4B08. [*Ham Radio License Manual*, page 5-7]

T4B10 **Which of the following is an appropriate receive filter bandwidth to select in order to minimize noise and interference for CW reception?**

A. 500 Hz
B. 1000 Hz
C. 2400 Hz
D. 5000 Hz

A See question T4B08. [*Ham Radio License Manual*, page 5-7]

T4B11 **Which of the following describes the common meaning of the term "repeater offset"?**

A. The distance between the repeater's transmit and receive antennas
B. The time delay before the repeater timer resets
C. The difference between the repeater's transmit and receive frequencies
D. Matching the antenna impedance to the feed line impedance

C To send a signal through a repeater, you must transmit on the repeater's input frequency where the repeater receiver listens, called the repeater's input frequency. You then listen where the repeater transmits, called the repeater's output frequency. To make it easy to use many different repeaters, hams have decided on a standard separation between input and output frequencies. The difference between repeater input and output frequencies is called the repeater's offset or shift. (See question T2A01.) [*Ham Radio License Manual*, page 6-16]

T4B12 What is the function of automatic gain control or AGC?

A. To keep received audio relatively constant
B. To protect an antenna from lightning
C. To eliminate RF on the station cabling
D. An asymmetric goniometer control used for antenna matching

A The AGC system in a receiver continually adjusts the sensitivity of the receiver so that the output audio volume is about the same to make listening more comfortable. AGC prevents strong signals from overdriving the output audio stages. Different modes (primarily CW and SSB) require different speeds at which the AGC system responds. [*Ham Radio License Manual*, page 5-7]

Electrical and Electronic Principles

SUBELEMENT T5 — Electrical principles: math for electronics; electronic principles; Ohm's Law [4 Exam Questions — 4 Groups]

T5A Electrical principles, units, and terms: current and voltage; conductors and insulators; alternating and direct current

T5A01 **Electrical current is measured in which of the following units?**

A. Volts
B. Watts
C. Ohms
D. Amperes

D The basic unit of electric current, a measure of the rate of flow of electrons, is the ampere, abbreviated A. (Current is represented by the symbol *I* or *i* in equations.) It is named for Andre Marie Ampere, an early 19th century scientist who studied electricity extensively. [*Ham Radio License Manual*, page 3-1]

T5A02 **Electrical power is measured in which of the following units?**

A. Volts
B. Watts
C. Ohms
D. Amperes

B Power is the rate at which energy is used. The basic unit of electrical power is the watt, abbreviated W. (Power is represented by the symbol *P* or *p* in equations.) This unit was named after James Watt, the 18th century inventor of the steam engine. [*Ham Radio License Manual*, page 3-5]

T5A03 What is the name for the flow of electrons in an electric circuit?

A. Voltage
B. Resistance
C. Capacitance
D. Current

D Electrons flow through the wires and components of an electric circuit. The flow of electrons in an electric circuit is called current. (See question T5A01.) [*Ham Radio License Manual*, page 3-1]

T5A04 What is the name for a current that flows only in one direction?

A. Alternating current
B. Direct current
C. Normal current
D. Smooth current

B Current that flows in one direction all the time is direct current, abbreviated dc. [*Ham Radio License Manual*, page 3-6]

T5A05 What is the electrical term for the electromotive force (EMF) that causes electron flow?

A. Voltage
B. Ampere-hours
C. Capacitance
D. Inductance

A Voltage is the electromotive force or electric potential that makes electrons move. Voltage is measured in volts, abbreviated as V. Voltage is represented in equations by the symbol E or e or v. [*Ham Radio License Manual*, page 3-1]

T5A06 How much voltage does a mobile transceiver usually require?

A. About 12 volts
B. About 30 volts
C. About 120 volts
D. About 240 volts

A Vehicle power systems are usually referred to as "12 volt", although the actual voltage typically varies from 12 to 15 V. Radios designed to operate from a "12 V" supply usually work best at the slightly higher voltage of 13.8 V, typical of vehicle power systems with the engine running. Check the equipment manual and be sure your power supply can generate the required voltage. [*Ham Radio License Manual*, page 5-15]

T5A07 Which of the following is a good electrical conductor?

A. Glass
B. Wood
C. Copper
D. Rubber

C In general, most metals make good conductors of electricity because their electrons are relatively free to move in response to an applied voltage. Copper, silver and aluminum are all excellent conductors, while wood, paper and glass are very poor conductors. [*Ham Radio License Manual*, page 3-4]

T5A08 Which of the following is a good electrical insulator?

A. Copper
B. Glass
C. Aluminum
D. Mercury

B Most non-metallic materials act as insulators because their electrons are not free to move. That means very little current will flow in response to an applied voltage. Ceramics, glass, paper and plastics are examples of good insulators. [*Ham Radio License Manual*, page 3-4]

T5A09 What is the name for a current that reverses direction on a regular basis?

A. Alternating current
B. Direct current
C. Circular current
D. Vertical current

A Alternating current, or ac, alternates direction, flowing first in one direction, then in the opposite direction. [*Ham Radio License Manual*, page 3-6]

T5A10 Which term describes the rate at which electrical energy is used?

A. Resistance
B. Current
C. Power
D. Voltage

C See question T5A02. [*Ham Radio License Manual*, page 3-5]

T5A11 What is the basic unit of electromotive force?

A. The volt
B. The watt
C. The ampere
D. The ohm

A See question T5A05. [*Ham Radio License Manual*, page 3-1]

T5A12 What term describes the number of times per second that an alternating current reverses direction?

A. Pulse rate
B. Speed
C. Wavelength
D. Frequency

D One complete sequence of alternating current (ac) flowing in one direction, stopping, reversing, and stopping again is a cycle. The number of cycles per second is the ac current's frequency. This also describes the frequency of a radio signal or wave. [*Ham Radio License Manual*, page 2-1]

T5B Math for electronics: conversion of electrical units; decibels; the metric system

T5B01 How many milliamperes is 1.5 amperes?

A. 15 milliamperes
B. 150 milliamperes
C. 1,500 milliamperes
D. 15,000 milliamperes

C The prefix "milli" means "divide by 1000," so "milliamperes" means "divide by 1000 to get amperes." 1500 milliamperes / 1000 = 1.5 amperes. Similarly, to convert from amperes to milliamperes, multiply by 1000, so 1.5 amperes × 1000 = 1500 milliamperes. [*Ham Radio License Manual*, page 2-2]

International System of Units (SI) — Metric Units

Prefix	*Symbol*	*Multiplication Factor*		
Tera	T	10^{12}	=	1,000,000,000,000
Giga	G	10^{9}	=	1,000,000,000
Mega	M	10^{6}	=	1,000,000
Kilo	k	10^{3}	=	1000
Hecto	h	10^{2}	=	100
Deca	da	10^{1}	=	10
Deci	d	10^{-1}	=	0.1
Centi	c	10^{-2}	=	0.01
Milli	m	10^{-3}	=	0.001
Micro	µ	10^{-6}	=	0.000001
Nano	n	10^{-9}	=	0.000000001
Pico	p	10^{-12}	=	0.000000000001

T5B02 What is another way to specify a radio signal frequency of 1,500,000 hertz?

A. 1500 kHz
B. 1500 MHz
C. 15 GHz
D. 150 kHz

A Checking each of the possible answers:

1500 kHz × 1000 = 1,500,000 Hz (the correct answer)

1500 MHz × 1,000,000 = 1,500,000,000 Hz (too large, 1,500,000 Hz = 1.5 MHz)

15 GHz × 1,000,000,000 = 1,500,000,000,000 Hz (very much too large, 1,500,000 Hz = 0.0015 GHz)

150 kHz × 1000 = 150,000 Hz (too small)

[*Ham Radio License Manual*, page 2-2]

T5B03 How many volts are equal to one kilovolt?

A. One one-thousandth of a volt
B. One hundred volts
C. One thousand volts
D. One million volts

C The prefix "kilo" means "multiply by 1000," so "kilovolts" means "multiply by 1000 to get volts." 1 kilovolt × 1000 = 1000 volts. [*Ham Radio License Manual*, page 2-2]

T5B04 How many volts are equal to one microvolt?

A. One one-millionth of a volt
B. One million volts
C. One thousand kilovolts
D. One one-thousandth of a volt

A The prefix "micro" means "divide by 1,000,000," so "microvolts" means "divide by 1,000,000 to get volts." 1 microvolt / 1,000,000 = 0.000001 V or one-millionth of a volt. [*Ham Radio License Manual*, page 2-2]

T5B05 Which of the following is equivalent to 500 milliwatts?

A. 0.02 watts
B. 0.5 watts
C. 5 watts
D. 50 watts

B The prefix "milli" means "divide by 1000," so "milliwatts" means "divide by 1000 to get watts." 500 milliwatts / 1000 = 0.5 W or watts. [*Ham Radio License Manual*, page 2-2]

T5B06 If an ammeter calibrated in amperes is used to measure a 3000-milliampere current, what reading would it show?

A. 0.003 amperes
B. 0.3 amperes
C. 3 amperes
D. 3,000,000 amperes

C The prefix "milli" means "divide by 1000," so "milliamperes" means "divide by 1000 to get amperes." 3000 milliamperes / 1000 = 3 A or amperes. [*Ham Radio License Manual*, page 2-2]

T5B07 If a frequency readout calibrated in megahertz shows a reading of 3.525 MHz, what would it show if it were calibrated in kilohertz?

A. 0.003525 kHz
B. 35.25 kHz
C. 3525 kHz
D. 3,525,000 kHz

C To convert from MHz to kHz, multiply by 1000 because 1000 kHz = 1 MHz. 3.525 × 1000 = 3525 kHz. [*Ham Radio License Manual*, page 2-2]

T5B08 How many microfarads are 1,000,000 picofarads?

A. 0.001 microfarads
B. 1 microfarad
C. 1000 microfarads
D. 1,000,000,000 microfarads

B To convert from picofarads to microfarads divide by 1,000,000, so 1,000,000 picofarads (or pF) = 1 microfarad (or μF). [*Ham Radio License Manual*, page 2-2]

T5B09 **What is the approximate amount of change, measured in decibels (dB), of a power increase from 5 watts to 10 watts?**

A. 2 dB
B. 3 dB
C. 5 dB
D. 10 dB

B Radio signals vary dramatically in strength. At the input to a receiver, signals are frequently smaller than one ten-billionth of a watt. When they come out of a transmitter, they're often measured in kilowatts! Electronic circuits change signal strengths by many factors of 10. These big differences in value make it difficult to compare signal sizes. Enter the decibel, abbreviated dB and pronounced "dee-bee." The decibel measures the ratio of two quantities as a power of 10. The formula for computing decibels is:

dB = 10 log (power ratio)

dB = 20 log (voltage ratio).

Positive values of dB mean the ratio is greater than 1 and negative values of dB indicate a ratio of less than 1. For example, if an amplifier turns a 5-watt signal into a 10-watt signal, that's a change of 10 log (10 / 5) = 10 log (2) = 3 dB. [*Ham Radio License Manual*, page 4-7]

T5B10 **What is the approximate amount of change, measured in decibels (dB), of a power decrease from 12 watts to 3 watts?**

A. –1 dB
B. –3 dB
C. –6 dB
D. –9 dB

C dB = 10 log (3 / 12) = 10 log (0.25) = –6 dB. [*Ham Radio License Manual*, page 4-7]

T5B11 **What is the approximate amount of change, measured in decibels (dB), of a power increase from 20 watts to 200 watts?**

A. 10 dB
B. 12 dB
C. 18 dB
D. 28 dB

A dB = 10 log (200 / 20) = 10 log (10) = 10 dB [*Ham Radio License Manual*, page 4-7]

T5B12 Which of the following frequencies is equal to 28,400 kHz?

A. 28.400 MHz
B. 2.800 MHz
C. 284.00 MHz
D. 28.400 kHz

A 1 MHz = 1000 kHz so 28,400 kHz / 1000 = 28.4 MHz. [*Ham Radio License Manual*, page 2-2]

T5B13 If a frequency readout shows a reading of 2425 MHz, what frequency is that in GHz?

A. 0.002425 GHz
B. 24.25 GHz
C. 2.425 GHz
D. 2425 GHz

C 1 GHz = 1000 MHz so 2425 MHz / 1000 = 2.425 GHz. [*Ham Radio License Manual*, page 2-2]

T5C Electronic principles: capacitance; inductance; current flow in circuits; alternating current; definition of RF; DC power calculations; impedance

T5C01 What is the ability to store energy in an electric field called?

A. Inductance
B. Resistance
C. Tolerance
D. Capacitance

D The ability to store energy in an electric field is called capacitance and it is measured in farads (F). Capacitors store electrical energy in the electric field created by a voltage between two conducting surfaces (such as metal foil) called electrodes. The electrodes are separated by an insulator called the dielectric. [*Ham Radio License Manual*, page 3-7]

T5C02 What is the basic unit of capacitance?

A. The farad
B. The ohm
C. The volt
D. The henry

A See question T5C01. [*Ham Radio License Manual*, page 3-7]

T5C03 What is the ability to store energy in a magnetic field called?

A. Admittance
B. Capacitance
C. Resistance
D. Inductance

D Inductors store magnetic energy in the magnetic field created by current flowing through a wire. This property is called inductance and it is measured in henrys (H). [*Ham Radio License Manual*, page 3-7]

T5C04 What is the basic unit of inductance?

A. The coulomb
B. The farad
C. The henry
D. The ohm

C See question T5C03. [*Ham Radio License Manual*, page 3-7]

T5C05 What is the unit of frequency?

A. Hertz
B. Henry
C. Farad
D. Tesla

A Each complete back-and-forth sequence of a radio signal (or any oscillating system, such as a pendulum) is called a cycle. The number of cycles per second is the signal's frequency, represented by a lower-case *f*. The unit of measurement for frequency is the hertz, abbreviated Hz. One cycle per second is one hertz or 1 Hz. As frequency increases, it becomes easier to use units of kilohertz (1 kHz = 1000 Hz), megahertz (1 MHz = 1000 kHz = 1,000,000 Hz), and gigahertz (1 GHz = 1000 MHz = 1,000,000,000 Hz). [*Ham Radio License Manual*, page 2-3]

T5C06 What does the abbreviation "RF" refer to?

A. Radio frequency signals of all types
B. The resonant frequency of a tuned circuit
C. The real frequency transmitted as opposed to the apparent frequency
D. Reflective force in antenna transmission lines

A Signals that have a frequency greater than 20,000 Hz (or 20 kHz) are radio frequency or RF signals. If connected to a speaker, signals below 20 kHz produce sound waves you can hear, so we call them audio frequency or AF signals. [*Ham Radio License Manual*, page 2-3]

T5C07 What is a usual name for electromagnetic waves that travel through space?

A. Gravity waves
B. Sound waves
C. Radio waves
D. Pressure waves

C A radio wave is a combination of an electric and a magnetic field, just like the fields in a capacitor and inductor but spreading out into space like ripples traveling across the surface of water. Because a radio wave is made up of both types of fields, it is an electromagnetic wave. (See questions T3A07 and T3B03.) [*Ham Radio License Manual*, page 4-6]

T5C08 What is the formula used to calculate electrical power in a DC circuit?

A. Power (P) equals voltage (E) multiplied by current (I)
B. Power (P) equals voltage (E) divided by current (I)
C. Power (P) equals voltage (E) minus current (I)
D. Power (P) equals voltage (E) plus current (I)

A Power is the product of voltage and current. As with Ohm's Law, if you know any two of P, E or I, you can determine the missing quantity as follows: $P = E \times I$ or $E = P / I$ or $I = P / E$. Voltage and current are never combined by addition or subtraction. [*Ham Radio License Manual*, page 3-5]

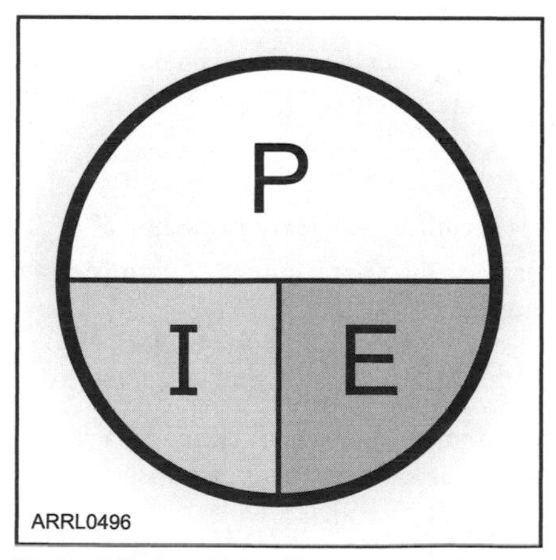

Figure T5-1 —This simple diagram will help you remember the power relationships. If you know any two of the quantities, you can find the third by covering up the unknown quantity. The positions of the remaining two symbols show if you have to multiply (side-by-side) or divide (one over the other).

T5C09 **How much power is being used in a circuit when the applied voltage is 13.8 volts DC and the current is 10 amperes?**

A. 138 watts
B. 0.7 watts
C. 23.8 watts
D. 3.8 watts

A Power = voltage × current = 13.8 V × 10 A = 138 W. [*Ham Radio License Manual*, page 3-5]

T5C10 **How much power is being used in a circuit when the applied voltage is 12 volts DC and the current is 2.5 amperes?**

A. 4.8 watts
B. 30 watts
C. 14.5 watts
D. 0.208 watts

B Power = voltage × current = 12 V × 2.5 A = 30 W. [*Ham Radio License Manual*, page 3-5]

T5C11 **How many amperes are flowing in a circuit when the applied voltage is 12 volts DC and the load is 120 watts?**

A. 0.1 amperes
B. 10 amperes
C. 12 amperes
D. 132 amperes

B From the equations presented in the explanation of question T5C08, current = power / voltage = 120 W / 12 V = 10 A. [*Ham Radio License Manual*, page 3-5]

T5C12 **What is meant by the term impedance?**

A. It is a measure of the opposition to AC current flow in a circuit
B. It is the inverse of resistance
C. It is a measure of the Q or Quality Factor of a component
D. It is a measure of the power handling capability of a component

A The combination of resistance and reactance is called impedance, represented by the capital letter Z, and is also measured in ohms (Ω). [*Ham Radio License Manual*, page 3-9]

T5C13 **What are the units of impedance?**

A. Volts
B. Amperes
C. Coulombs
D. Ohms

D See question T5C12. [*Ham Radio License Manual*, page 3-9]

T5D Ohm's Law: formulas and usage

T5D01 What formula is used to calculate current in a circuit?

A. Current (I) equals voltage (E) multiplied by resistance (R)
B. Current (I) equals voltage (E) divided by resistance (R)
C. Current (I) equals voltage (E) added to resistance (R)
D. Current (I) equals voltage (E) minus resistance (R)

B Ohm's Law states that $R = E / I$. If you know any two of *I*, *E*, or *R*, you can determine the missing quantity as follows: $R = E / I$, $I = E / R$, or $E = I \times R$. The drawing in the figure below is a convenient aid to remembering Ohm's Law in any of these three forms. [*Ham Radio License Manual*, page 3-4]

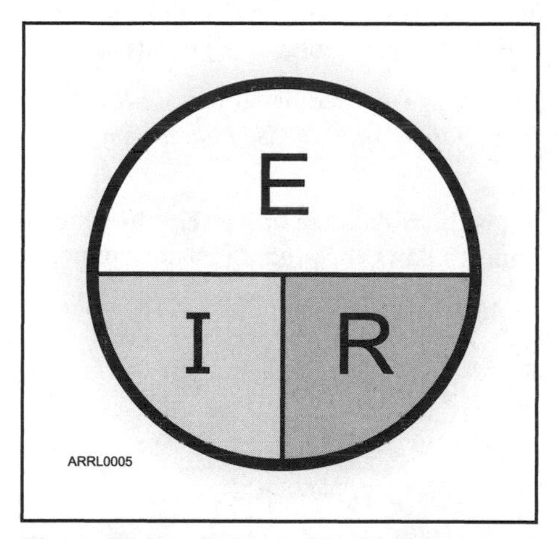

Figure T5-2 — This simple diagram will help you remember the Ohm's Law relationships. If you know any two of the quantities, you can find the third by covering up the unknown quantity. The positions of the remaining two symbols show if you have to multiply (side-by-side) or divide (one above the other).

T5D02 **What formula is used to calculate voltage in a circuit?**

A. Voltage (E) equals current (I) multiplied by resistance (R)
B. Voltage (E) equals current (I) divided by resistance (R)
C. Voltage (E) equals current (I) added to resistance (R)
D. Voltage (E) equals current (I) minus resistance (R)

A Voltage = current × resistance, or $E = I \times R$. (See question T5D01.) [*Ham Radio License Manual*, page 3-4]

T5D03 **What formula is used to calculate resistance in a circuit?**

A. Resistance (R) equals voltage (E) multiplied by current (I)
B. Resistance (R) equals voltage (E) divided by current (I)
C. Resistance (R) equals voltage (E) added to current (I)
D. Resistance (R) equals voltage (E) minus current (I)

B Resistance = voltage divided by current, or $R = E / I$. (See question T5D01.) [*Ham Radio License Manual*, page 3-4]

T5D04 **What is the resistance of a circuit in which a current of 3 amperes flows through a resistor connected to 90 volts?**

A. 3 ohms
B. 30 ohms
C. 93 ohms
D. 270 ohms

B $R = E / I = 90$ V / 3 A = 30 Ω or ohms. (See question T5D01.) [*Ham Radio License Manual*, page 3-5]

T5D05 **What is the resistance in a circuit for which the applied voltage is 12 volts and the current flow is 1.5 amperes?**

A. 18 ohms
B. 0.125 ohms
C. 8 ohms
D. 13.5 ohms

C $R = E / I = 12$ V / 1.5 A = 8 Ω or ohms. (See question T5D01.) [*Ham Radio License Manual*, page 3-5]

T5D06 **What is the resistance of a circuit that draws 4 amperes from a 12-volt source?**

A. 3 ohms
B. 16 ohms
C. 48 ohms
D. 8 ohms

A $R = E / I = 12$ V / 4 A = 3 Ω or ohms (See question T5D01.) [*Ham Radio License Manual*, page 3-5]

T5D07 **What is the current flow in a circuit with an applied voltage of 120 volts and a resistance of 80 ohms?**

A. 9600 amperes
B. 200 amperes
C. 0.667 amperes
D. 1.5 amperes

D $I = E / R = 120 \text{ V} / 80\ \Omega = 1.5 \text{ A}$ or amperes. (See question T5D01.) [*Ham Radio License Manual*, page 3-5]

T5D08 **What is the current flowing through a 100-ohm resistor connected across 200 volts?**

A. 20,000 amperes
B. 0.5 amperes
C. 2 amperes
D. 100 amperes

C $I = E / R = 200 \text{ V} / 100\ \Omega = 2 \text{ A}$ or amperes. (See question T5D01.) [*Ham Radio License Manual*, page 3-5]

T5D09 **What is the current flowing through a 24-ohm resistor connected across 240 volts?**

A. 24,000 amperes
B. 0.1 amperes
C. 10 amperes
D. 216 amperes

C $I = E / R = 240 \text{ V} / 24\ \Omega = 10 \text{ A}$ or amperes. (See question T5D01) [*Ham Radio License Manual*, page 3-5]

T5D10 **What is the voltage across a 2-ohm resistor if a current of 0.5 amperes flows through it?**

A. 1 volt
B. 0.25 volts
C. 2.5 volts
D. 1.5 volts

A $E = I \times R = 0.5 \text{ A} \times 2\ \Omega = 1 \text{ V}$ or volt. (See question T5D01.) [*Ham Radio License Manual*, page 3-5]

T5D11 **What is the voltage across a 10-ohm resistor if a current of 1 ampere flows through it?**

A. 1 volt
B. 10 volts
C. 11 volts
D. 9 volts

B $E = I \times R = 1\ \text{A} \times 10\ \Omega = 10\ \text{V}$ or volts. (See question T5D01.) [*Ham Radio License Manual*, page 3-5]

T5D12 **What is the voltage across a 10-ohm resistor if a current of 2 amperes flows through it?**

A. 8 volts
B. 0.2 volts
C. 12 volts
D. 20 volts

D $E = I \times R = 2\ \text{A} \times 10\ \Omega = 20\ \text{V}$ or volts. (See question T5D01.) [*Ham Radio License Manual*, page 3-5]

Electrical Components and Functions

SUBELEMENT T6 — Electrical components: semiconductors; circuit diagrams; component functions [4 Exam Questions — 4 Groups]

T6A Electrical components: fixed and variable resistors; capacitors and inductors; fuses; switches; batteries

T6A01 What electrical component is used to oppose the flow of current in a DC circuit?

A. Inductor
B. Resistor
C. Voltmeter
D. Transformer

B Resistors have a certain resistance specified in ohms (Ω), kilo-ohms (kΩ), or mega-ohms (MΩ). The function of a resistor is to oppose the flow of electrical current, just as a valve in a water pipe restricts the flow through the pipe. [*Ham Radio License Manual*, page 3-7]

T6A02 What type of component is often used as an adjustable volume control?

A. Fixed resistor
B. Power resistor
C. Potentiometer
D. Transformer

C A variable resistor is also called a potentiometer (poh-ten-chee-AH-meh-tur) or pot because it is frequently used to adjust voltage or potential, such as for a volume control. [*Ham Radio License Manual*, page 3-9]

T6A03 What electrical parameter is controlled by a potentiometer?

A. Inductance
B. Resistance
C. Capacitance
D. Field strength

B See question T6A02. [*Ham Radio License Manual*, page 3-9]

T6A04 What electrical component stores energy in an electric field?

A. Resistor
B. Capacitor
C. Inductor
D. Diode

B Capacitors store electrical energy in the electric field created by a voltage between two conducting surfaces (such as metal foil) called electrodes. The electrodes are separated by an insulator called the dielectric. (See question T5C01.) [*Ham Radio License Manual*, page 3-7]

T6A05 What type of electrical component consists of two or more conductive surfaces separated by an insulator?

A. Resistor
B. Potentiometer
C. Oscillator
D. Capacitor

D See question T6A04. [*Ham Radio License Manual*, page 3-7]

T6A06 What type of electrical component stores energy in a magnetic field?

A. Resistor
B. Capacitor
C. Inductor
D. Diode

C Inductors store magnetic energy in the magnetic field created by current flowing through a wire. This is called inductance and it is measured in henrys (H). Inductors are made from wire wound in a coil, sometimes around a core of magnetic material that concentrates the magnetic energy. (See question T5C03.) [*Ham Radio License Manual*, page 3-7]

T6A07 What electrical component is usually composed of a coil of wire?

A. Switch
B. Capacitor
C. Diode
D. Inductor

D See question T6A06. [*Ham Radio License Manual*, page 3-7]

T6A08 What electrical component is used to connect or disconnect electrical circuits?

A. Magnetron
B. Switch
C. Thermistor
D. All of these choices are correct

B Switches and relays are simple components that control current through a circuit by connecting and disconnecting the paths current can follow. Both can interrupt current — called opening a circuit — or allow it to flow — called closing a circuit. [*Ham Radio License Manual*, page 3-12]

T6A09 What electrical component is used to protect other circuit components from current overloads?

A. Fuse
B. Capacitor
C. Inductor
D. All of these choices are correct

A Fuses interrupt current overloads by melting a short length of metal. When the metal melts or "blows," the current path is broken and power is removed from circuits supplied by the fuse. [*Ham Radio License Manual*, page 3-12]

T6A10 Which of the following battery types is rechargeable?

A. Nickel-metal hydride
B. Lithium-ion
C. Lead-acid gel-cell
D. All of these choices are correct

D Secondary or rechargeable batteries are widely used in Amateur Radio. Whether a battery is rechargeable or not depends on its chemistry — the type of chemical reaction by which the battery stores and delivers electrical energy. The table lists the chemistries and electrical characteristics of batteries commonly used by hams. [*Ham Radio License Manual*, page 5-17]

Battery Types and Characteristics

Battery Style	*Chemistry Type*	*Fully-Charged Voltage*	*Energy Rating (average)*
AAA	Alkaline — Disposable	1.5 V	1100 mAh
AA	Alkaline — Disposable	1.5 V	2600 – 3200 mAh
AA	Carbon-Zinc — Disposable	1.5 V	600 mAh
AA	Nickel-Cadmium (NiCd) — Rechargeable	1.2 V	700 mAh
AA	Nickel-Metal Hydride (NiMH) — Rechargeable	1.2 V	1500 – 2200 mAh
C	Alkaline — Disposable	1.5 V	7500 mAh
D	Alkaline — Disposable	1.5 V	14000 mAh
9 V	Alkaline — Disposable	9 V	580 mAh
9 V	Nickel-Cadmium (NiCd) — Rechargeable	9 V	110 mAh
9 V	Nickel-Metal Hydride — Rechargeable	9 V	150 mAh
Coin Cells	Lithium — Disposable	3 – 3.3 V	25 – 1000 mA
Packs	Lithium Ion — Rechargeable	3.3 – 3.6 V per cell	

T6A11 Which battery type is not rechargeable?

A. Nickel-cadmium
B. Carbon-zinc
C. Lead-acid
D. Lithium-ion

B Carbon-zinc batteries are disposable, not rechargeable, as shown in the table of battery types and characteristics. [*Ham Radio License Manual*, page 5-17]

T6B Semiconductors: basic principles and applications of solid state devices; diodes and transistors

T6B01 What class of electronic components is capable of using a voltage or current signal to control current flow?

A. Capacitors
B. Inductors
C. Resistors
D. Transistors

D Transistors are components made from patterns of N- and P-type semiconductor material creating by doping — the adding of impurities to change the way the material conducts electricity. The patterns form structures that allow the transistor to use small voltages and currents to control larger ones. The transistor's electrodes are contacts made to a separate piece of the pattern. With the appropriate external circuit and a source of power, transistors can amplify or switch voltages and currents. Using small signals to control larger signals is called gain. [*Ham Radio License Manual*, page 3-11]

T6B02 What electronic component allows current to flow in only one direction?

A. Resistor
B. Fuse
C. Diode
D. Driven Element

C When N- and P-type material are placed in contact with each other, the result is a PN junction that conducts better in one direction than the other. This and other properties are used to create many useful electronic components, generally referred to as semiconductors. A semiconductor that only allows current flow in one direction is called a diode. Current flows from the anode to the cathode. [*Ham Radio License Manual*, page 3-10]

T6B03 Which of these components can be used as an electronic switch or amplifier?

A. Oscillator
B. Potentiometer
C. Transistor
D. Voltmeter

C See question T6B01. [*Ham Radio License Manual*, page 3-11]

T6B04 Which of the following components can be made of three layers of semiconductor material?

A. Alternator
B. Transistor
C. Triode
D. Pentagrid converter

B The bipolar junction transistor (BJT) is made from three layers of alternating N-type and P-type material that form the base, emitter, and collector electrodes. (See also question T6B01.) [*Ham Radio License Manual*, page 3-11]

T6B05 Which of the following electronic components can amplify signals?

A. Transistor
B. Variable resistor
C. Electrolytic capacitor
D. Multi-cell battery

A See question T6B01. [*Ham Radio License Manual*, page 3-11]

T6B06 How is the cathode lead of a semiconductor diode usually identified?

A. With the word cathode
B. With a stripe
C. With the letter C
D. All of these choices are correct

B Most semiconductor diode components are made with a protective coating around the actual semiconductor material. The electrodes are a pair of wire leads or tabs that make contact to the semiconductor material. Short of actually performing an electrical test, there is no way to tell the electrodes apart except for marking or labeling. The most common method is marking the body of the diode with a paint stripe to indicate the cathode electrode. [*Ham Radio License Manual*, page 3-10]

T6B07 What does the abbreviation LED stand for?

A. Low Emission Diode
B. Light Emitting Diode
C. Liquid Emission Detector
D. Long Echo Delay

B The light-emitting diode or LED, gives off light when current flows through it. [*Ham Radio License Manual*, page 3-10]

T6B08 What does the abbreviation FET stand for?

A. Field Effect Transistor
B. Fast Electron Transistor
C. Free Electron Transition
D. Field Emission Thickness

A The FET (or field effect transistor) is constructed as a conducting path or channel of N- or P-type material. The ends of the channel form the source and drain electrodes. The gate electrode is used to control current flow through the channel. [*Ham Radio License Manual*, page 3-11]

T6B09 What are the names of the two electrodes of a diode?

A. Plus and minus
B. Source and drain
C. Anode and cathode
D. Gate and base

C See question T6B02. [*Ham Radio License Manual*, page 3-10]

T6B10 What are the three electrodes of a PNP or NPN transistor?

A. Emitter, base, and collector
B. Source, gate, and drain
C. Cathode, grid, and plate
D. Cathode, drift cavity, and collector

A A transistor is generally fabricated from a single piece of semiconductor material with patterns or layers of different types of material. Each layer or piece of the pattern that is connected to an external terminal is called an electrode. Bipolar junction transistors (BJT) have three electrodes: emitter, base, and collector. [*Ham Radio License Manual*, page 3-11]

T6B11 What are the three electrodes of a field effect transistor?

A. Emitter, base, and collector
B. Source, gate, and drain
C. Cathode, grid, and plate
D. Cathode, gate, and anode

B The electrodes of the field effect transistor (FET) are the source, gate, and drain. (See question T6B08.) [*Ham Radio License Manual*, page 3-11]

T6B12 What is the term that describes a transistor's ability to amplify a signal?

A. Gain
B. Forward resistance
C. Forward voltage drop
D. On resistance

A See question T6B01. [*Ham Radio License Manual*, page 3-11]

T6C Circuit diagrams; schematic symbols

T6C01 What is the name for standardized representations of components in an electrical wiring diagram?

A. Electrical depictions
B. Grey sketch
C. Schematic symbols
D. Component callouts

C If a circuit contains more than two or three components, trying to describe it clearly in words becomes very difficult. To describe complicated circuits, engineers have developed the *s*chematic diagram or simply schematic. Schematics create a visual description of a circuit by using standardized representations of electrical components called circuit symbols. Figure T6-1 shows commonly used schematic symbols. [*Ham Radio License Manual*, page 3-13]

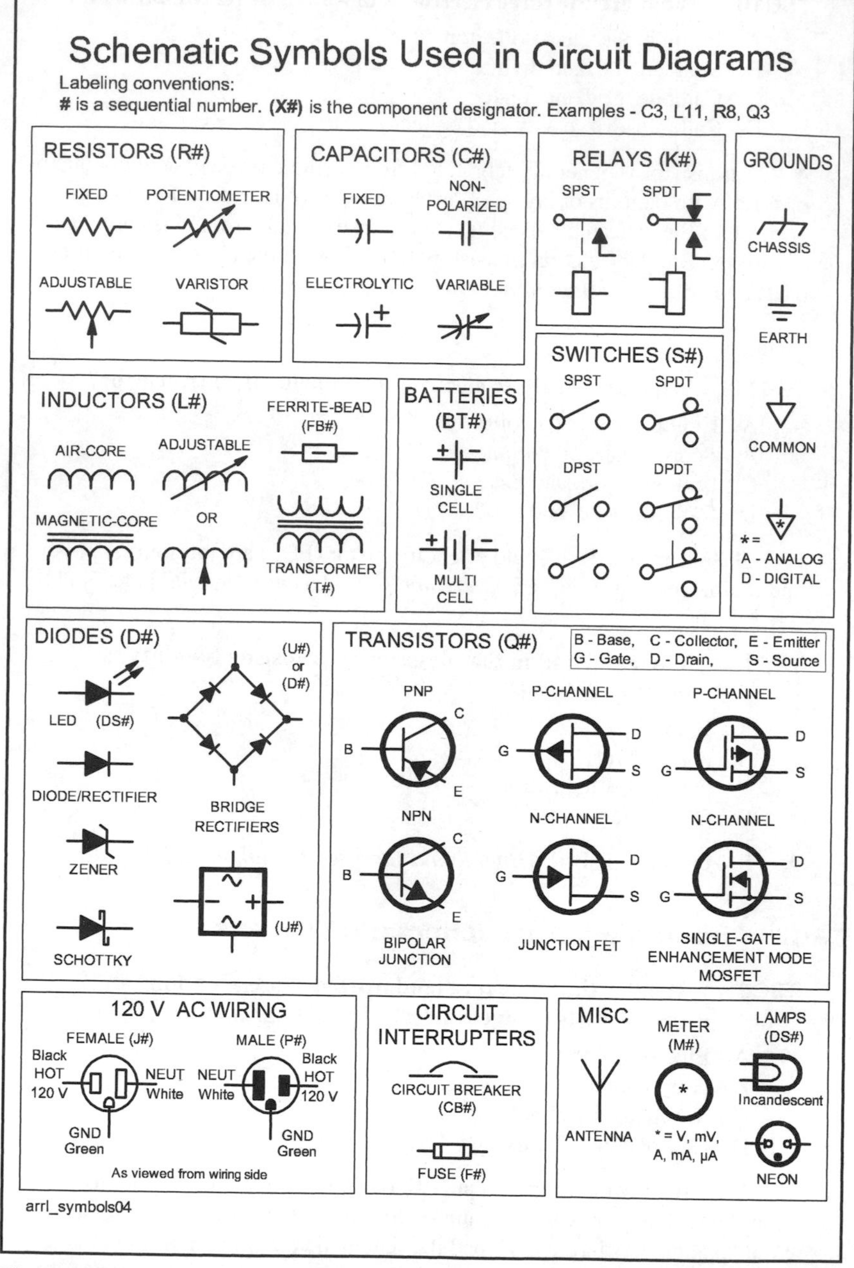

Figure T6-1 — Questions in the following sections ask you to identify specific schematic symbols, and you can use this figure to identify each of those symbols. Circuits T1, T2 and T3 are constructed entirely from symbols defined in this figure.

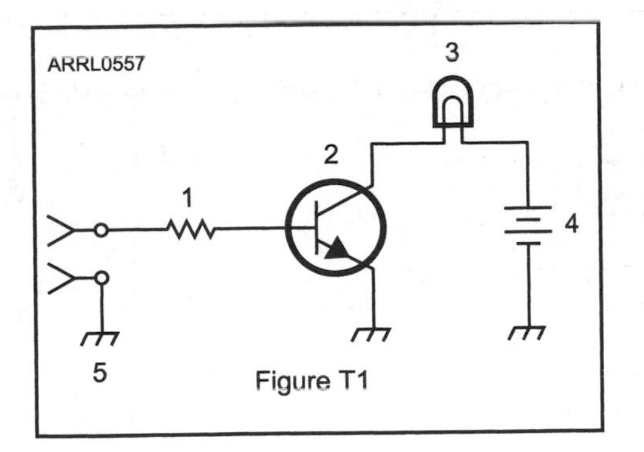

Figure T1

T6C02 What is component 1 in figure T1?

A. Resistor
B. Transistor
C. Battery
D. Connector

A See Figure T6-1. [*Ham Radio License Manual*, page 3-13]

T6C03 What is component 2 in figure T1?

A. Resistor
B. Transistor
C. Indicator lamp
D. Connector

B See Figure T6-1. [*Ham Radio License Manual*, page 3-13]

T6C04 What is component 3 in figure T1?

A. Resistor
B. Transistor
C. Lamp
D. Ground symbol

C See Figure T6-1. [*Ham Radio License Manual*, page 3-13]

T6C05 What is component 4 in figure T1?

A. Resistor
B. Transistor
C. Battery
D. Ground symbol

C See Figure T6-1. [*Ham Radio License Manual*, page 3-13]

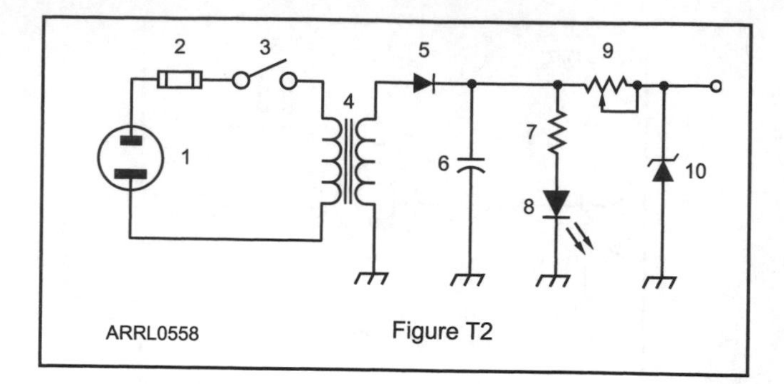

Figure T2

T6C06 What is component 6 in figure T2?

A. Resistor
B. Capacitor
C. Regulator IC
D. Transistor

B See Figure T6-1. [*Ham Radio License Manual*, page 3-13]

T6C07 What is component 8 in figure T2?

A. Resistor
B. Inductor
C. Regulator IC
D. Light emitting diode

D See Figure T6-1. [*Ham Radio License Manual*, page 3-13]

T6C08 What is component 9 in figure T2?

A. Variable capacitor
B. Variable inductor
C. Variable resistor
D. Variable transformer

C See Figure T6-1. [*Ham Radio License Manual*, page 3-13]

T6C09 What is component 4 in figure T2?

A. Variable inductor
B. Double-pole switch
C. Potentiometer
D. Transformer

D See Figure T6-1. [*Ham Radio License Manual*, page 3-13]

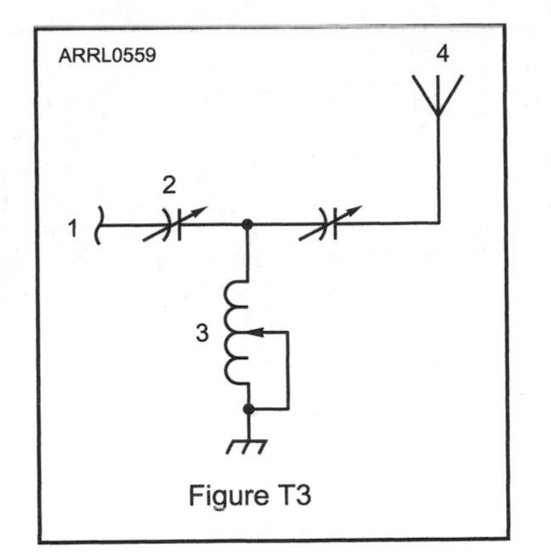

Figure T3

T6C10 What is component 3 in figure T3?

A. Connector
B. Meter
C. Variable capacitor
D. Variable inductor

D See Figure T6-1. [*Ham Radio License Manual*, page 3-13]

T6C11 What is component 4 in figure T3?

A. Antenna
B. Transmitter
C. Dummy load
D. Ground

A See Figure T6-1. [*Ham Radio License Manual*, page 3-13]

T6C12 What do the symbols on an electrical circuit schematic diagram represent?

A. Electrical components
B. Logic states
C. Digital codes
D. Traffic nodes

A See question T6C01. [*Ham Radio License Manual*, page 3-13]

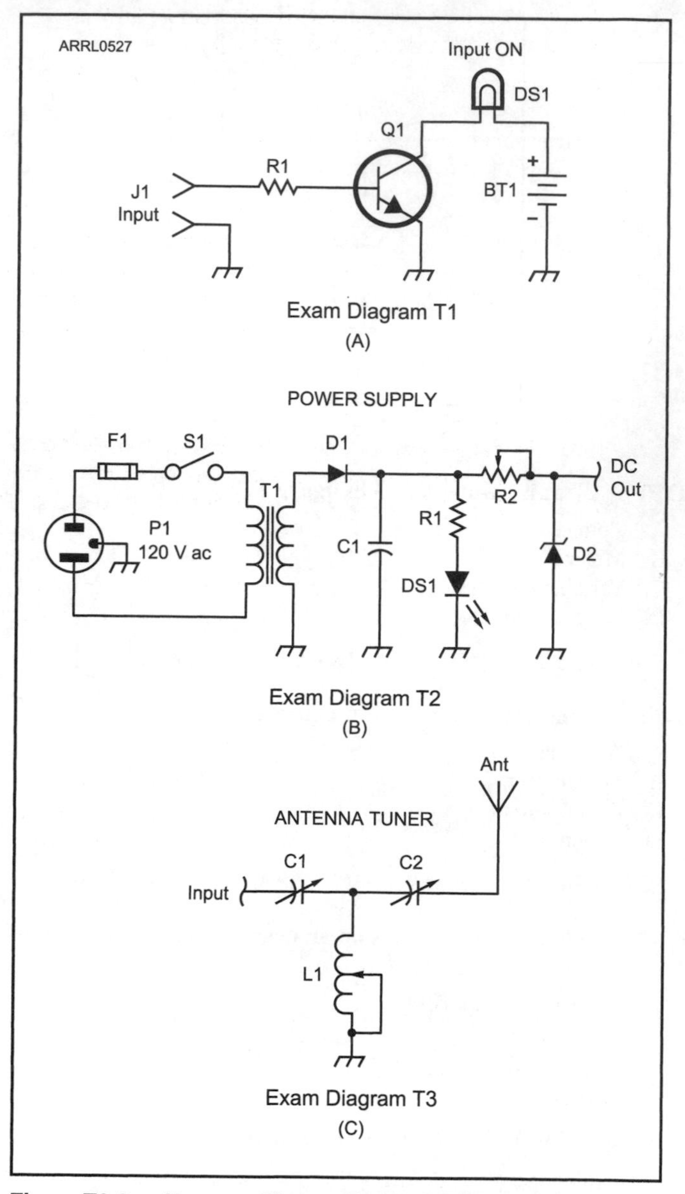

Figure T6-2 — Here are Figures T1 (A), T2 (B) and T3 (C) used on the Technician exam with component designators drawn in. Different letters represent each type of component, for example R for resistor (see Figure T6-1). Lines and dots show electrical connections between the components, but may not correspond to actual wires.

T6C13 Which of the following is accurately represented in electrical circuit schematic diagrams?

A. Wire lengths
B. Physical appearance of components
C. The way components are interconnected
D. All of these choices are correct

C A schematic does not illustrate the actual physical layout of a circuit. (A pictorial diagram is used for that purpose.) It only shows how the components are connected electrically. The lines drawn from component to component represent those electrical connections. Each line does not necessarily correspond to a physical wire — it just indicates that an electrical connection exists between whatever is at each end of the line. [*Ham Radio License Manual*, page 3-14]

T6D Component functions: rectification; switches; indicators; power supply components; resonant circuit; shielding; power transformers; integrated circuits

T6D01 Which of the following devices or circuits changes an alternating current into a varying direct current signal?

A. Transformer
B. Rectifier
C. Amplifier
D. Reflector

B Heavy-duty diodes that can withstand large voltages and currents are called rectifiers. If an ac voltage is applied to a diode, the result is a unidirectional, pulsing dc current because current is blocked when the voltage tries to push electrons in the "wrong" direction. (See question T6B02.) Circuits that change ac current into dc are also called rectifiers. [*Ham Radio License Manual*, page 3-10]

T6D02 What best describes a relay?

A. A switch controlled by an electromagnet
B. A current controlled amplifier
C. An optical sensor
D. A pass transistor

A A switch is operated manually while a relay is a switch controlled by an electromagnet. [*Ham Radio License Manual*, page 3-12]

T6D03 **What type of switch is represented by component 3 in figure T2?**

A. Single-pole single-throw
B. Single-pole double-throw
C. Double-pole single-throw
D. Double-pole double-throw

A Figure T2 is shown with question T6C06. Use the symbol chart shown in Figure T6-1 as for questions T6C02 through T6C11. [*Ham Radio License Manual*, page 3-13]

T6D04 **Which of the following can be used to display signal strength on a numeric scale?**

A. Potentiometer
B. Transistor
C. Meter
D. Relay

C Indicators and displays are important components for radio equipment. An indicator is either ON or OFF, such as a power indicator or a label that appears when you are transmitting. A meter provides information as a value in the form of numbers or on a numeric scale. [*Ham Radio License Manual*, page 3-13]

T6D05 **What type of circuit controls the amount of voltage from a power supply?**

A. Regulator
B. Oscillator
C. Filter
D. Phase inverter

A A power supply's output voltage changes with the amount of output current. The percentage of voltage change between zero current (no load) and maximum current (full load) is the regulation of the supply. To achieve "tight" regulation, meaning little variation as current changes, requires a regulator circuit in the supply. [*Ham Radio License Manual*, page 5-15]

T6D06 What component is commonly used to change 120V AC house current to a lower AC voltage for other uses?

A. Variable capacitor
B. Transformer
C. Transistor
D. Diode

B Transformers are made from two or more inductors that share their stored energy. This allows energy to be transferred from one inductor to another while changing the combination of voltage and current. A transformer is used to transfer energy from a home's 120 V ac outlets to a lower voltage for use in electronic equipment. [*Ham Radio License Manual*, page 3-9]

T6D07 Which of the following is commonly used as a visual indicator?

A. LED
B. FET
C. Zener diode
D. Bipolar transistor

A The light emitting diode (LED) is often used as an indicator on electronic equipment. See questions T6B07 and T6D04. [*Ham Radio License Manual*, page 3-11]

T6D08 Which of the following is used together with an inductor to make a tuned circuit?

A. Resistor
B. Zener diode
C. Potentiometer
D. Capacitor

D Circuits that contain both capacitors and inductors will have at least one resonant frequency and are called resonant circuits or tuned circuits. Capacitors and inductors are used to create tuned circuits. A tuned circuit acts as a filter, passing or rejecting signals at its resonant frequency. Tuned circuits are important in radio because they help radios generate or receive signals at a single frequency. [*Ham Radio License Manual*, page 3-9]

T6D09 **What is the name of a device that combines several semiconductors and other components into one package?**

A. Transducer
B. Multi-pole relay
C. Integrated circuit
D. Transformer

C An integrated circuit (IC or chip) is made of many components connected together as a useful circuit and packaged as a single component. ICs range from very simple circuits consisting of a few diodes all the way to complex microprocessors or signal-processing chips with many thousands of components. [*Ham Radio License Manual*, page 3-11]

T6D10 **What is the function of component 2 in Figure T1?**

A. Give off light when current flows through it
B. Supply electrical energy
C. Control the flow of current
D. Convert electrical energy into radio waves

C Figure T1 is shown with question T6C06. Component 2 is a transistor that can act as a switch or amplifier. (See question T6B01 and Figure T6-1.) [*Ham Radio License Manual*, page 3-11]

T6D11 **What is a simple resonant or tuned circuit?**

A. An inductor and a capacitor connected in series or parallel to form a filter
B. A type of voltage regulator
C. A resistor circuit used for reducing standing wave ratio
D. A circuit designed to provide high fidelity audio

A See question T6D08. [*Ham Radio License Manual*, page 3-9]

T6D12 **Which of the following is a common reason to use shielded wire?**

A. To decrease the resistance of DC power connections
B. To increase the current carrying capability of the wire
C. To prevent coupling of unwanted signals to or from the wire
D. To couple the wire to other signals

C The shield surrounding the inner wire acts as a barrier, keeping unwanted signals from being picked up by the wire and preventing signals flowing in the wire from radiating. [*Ham Radio License Manual*, page 5-22]

Station Equipment

SUBELEMENT T7 — Station equipment: common transmitter and receiver problems; antenna measurements; troubleshooting; basic repair and testing [4 Exam Questions — 4 Groups]

T7A Station equipment: receivers; transmitters; transceivers; modulation; transverters; low power and weak signal operation; transmit and receive amplifiers

T7A01 Which term describes the ability of a receiver to detect the presence of a signal?

A. Linearity
B. Sensitivity
C. Selectivity
D. Total Harmonic Distortion

B Receivers are compared on the basis of two primary characteristics: sensitivity and selectivity. A receiver's sensitivity determines its ability to detect weak signals. Higher sensitivity means a receiver can detect weaker signals. Sensitivity is specified as a minimum discernable signal or minimum detectable signal level, usually in microvolts. Selectivity is the ability of a receiver to retrieve the information from just the desired signal in the presence of unwanted signals. High selectivity means that a receiver can operate properly even in the presence of strong signals on nearby frequencies. Receivers use filters to reject the unwanted signals. [*Ham Radio License Manual*, page 3-18]

T7A02 What is a transceiver?

A. A type of antenna switch
B. A unit combining the functions of a transmitter and a receiver
C. A component in a repeater which filters out unwanted interference
D. A type of antenna matching network

B Most radios today are transceivers although home-made equipment might still consist of a separate transmitter and receiver. In a transceiver, a transmit-receive switch — a relay or an electronic circuit — allows both the transmit and receive circuits to share a common antenna. [*Ham Radio License Manual*, page 2-12]

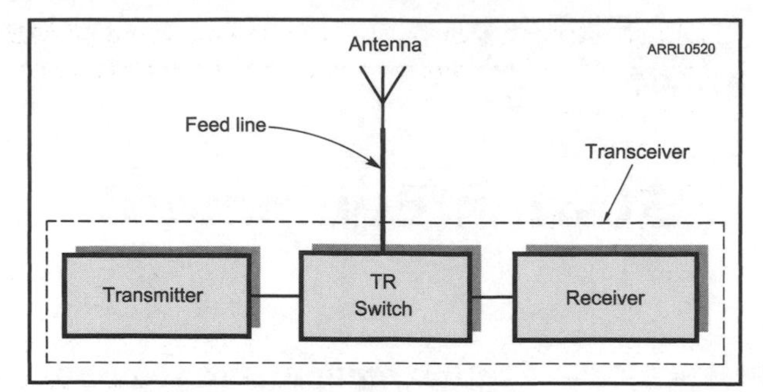

Figure T7-1 — A transceiver consists of a transmitter and receiver in a single package. The transmit-receive (TR) switch allows the transmitter and receiver to share a single antenna.

T7A03 Which of the following is used to convert a radio signal from one frequency to another?

A. Phase splitter
B. Mixer
C. Inverter
D. Amplifier

B A mixer combines two signals and creates output signals called mixing products at the sum and difference frequencies of the input signals. [*Ham Radio License Manual*, page 3-18]

T7A04 Which term describes the ability of a receiver to discriminate between multiple signals?

A. Discrimination ratio
B. Sensitivity
C. Selectivity
D. Harmonic Distortion

C See question T7A01. [*Ham Radio License Manual*, page 3-18]

T7A05 What is the name of a circuit that generates a signal of a desired frequency?

A. Reactance modulator
B. Product detector
C. Low-pass filter
D. Oscillator

D Most oscillators used in radio equipment produce a single-frequency sine wave but other types of oscillators produce square, triangle, sawtooth, pulse and other waveforms. [*Ham Radio License Manual*, page 3-16]

T7A06 What device takes the output of a low-powered 28 MHz SSB exciter and produces a 222 MHz output signal?

A. High-pass filter
B. Low-pass filter
C. Transverter
D. Phase converter

C Low-power transceiver RF output signals on one band are shifted by a transverter to the new output frequency where they are amplified for transmission. A receiving converter in the transverter shifts incoming signals to the desired band where they are received as regular signals by the transceiver. Transverters are used by many hams to allow one main transceiver to be used on one or more new bands. For example, with few transceivers available for CW and SSB operation on 222 MHz, a transverter is used to convert 222 MHz signals to and from the 28 MHz band available on all HF gear. [*Ham Radio License Manual*, page 3-19]

T7A07 What is meant by the term "PTT"?

A. Pre-transmission tuning to reduce transmitter harmonic emission
B. Precise tone transmissions used to limit repeater access to only certain signals
C. A primary transformer tuner use to match antennas
D. The push to talk function which switches between receive and transmit

D The PTT switch is located on the microphone or on the side of a hand-held transceiver. [*Ham Radio License Manual*, page 5-6]

T7A08 Which of the following describes combining speech with an RF carrier signal?

A. Impedance matching
B. Oscillation
C. Modulation
D. Low-pass filtering

C The process of combining data or voice signals with an RF signal is modulation. A circuit that performs the modulation function is therefore called a modulator. The function of the modulator is to combine the data or voice signal with an RF signal or carrier. The result is an RF signal that can be communicated by radio. A modulator can be as simple as an on-off switch (a telegraph key, for example) or it can be very complex in the case of a high-speed data transmission. [*Ham Radio License Manual*, page 3-17]

T7A09 Which of the following devices is most useful for VHF weak-signal communication?

A. A quarter-wave vertical antenna
B. A multi-mode VHF transceiver
C. An omni-directional antenna
D. A mobile VHF FM transceiver

B Most VHF "weak-signal" contacts are made using SSB or CW or specialized digital modes because those modes give better performance over long distances. This style of operating also uses horizontally-polarized directional antennas (beams). Since FM is not used, you'll need a multi-mode transceiver to use SSB or CW or digital modes for VHF/UHF DXing. [*Ham Radio License Manual*, page 6-28]

T7A10 What device increases the low-power output from a handheld transceiver?

A. A voltage divider
B. An RF power amplifier
C. An impedance network
D. All of these choices are correct

B Amplifiers are used to increase the strength of audio or radio signals. Amplifiers used to increase the power of transmitted RF signals are called power amplifiers or linear amplifiers. Amplifiers used to increase the strength of a received signal are called preamplifiers or just preamps. [*Ham Radio License Manual*, page 5-8]

T7A11 Where is an RF preamplifier installed?

A. Between the antenna and receiver
B. At the output of the transmitter's power amplifier
C. Between a transmitter and antenna tuner
D. At the receiver's audio output

A If a receiver is not sensitive enough (see question T7A01), a preamplifier (or preamp) can be connected between the antenna and receiver. [*Ham Radio License Manual*, page 3-18]

T7B Common transmitter and receiver problems: symptoms of overload and overdrive; distortion; causes of interference; interference and consumer electronics; part 15 devices; over and under modulation; RF feedback; off frequency signals; fading and noise; problems with digital communications interfaces

T7B01 What can you do if you are told your FM handheld or mobile transceiver is over-deviating?

A. Talk louder into the microphone
B. Let the transceiver cool off
C. Change to a higher power level
D. Talk farther away from the microphone

D Excessive modulation for all types of speech results in distortion of transmitted speech and unwanted or spurious transmitter outputs on adjacent frequencies where they cause interference. Those unwanted transmitter outputs have lots of names, but the most common is splatter. Generating those outputs is called splattering, as in, "You're splattering 10 kHz away!" An overmodulated FM signal has excessive deviation and is said to be over-deviating. Over-deviation is usually caused by speaking too loudly into the microphone and may cause interference on adjacent channels. An FM transmitter can also be internally misadjusted to over-deviate at normal speech levels. To reduce over-deviation, speak more softly or move the microphone farther from your mouth. [*Ham Radio License Manual*, page 5-4]

T7B02 What would cause a broadcast AM or FM radio to receive an amateur radio transmission unintentionally?

A. The receiver is unable to reject strong signals outside the AM or FM band
B. The microphone gain of the transmitter is turned up too high
C. The audio amplifier of the transmitter is overloaded
D. The deviation of an FM transmitter is set too low

A Consumer-type broadcast receivers are not designed to reject extremely strong signals outside the broadcast bands. A nearby amateur station may produce a signal strong enough to overload the receiver or be picked up by the receiver circuitry directly. [*Ham Radio License Manual*, page 5-21]

T7B03 **Which of the following may be a cause of radio frequency interference?**

A. Fundamental overload
B. Harmonics
C. Spurious emissions
D. All of these choices are correct

D All three of the choices cause interference by disrupting receiver performance (A) or by creating unwanted signals that directly interfere with desired signals (B and C). As more and more electronic devices and electrical appliances are put in use every day, interference between them and radios, called radio frequency interference (RFI), becomes more commonplace. RFI can occur in either "direction" — to or from the Amateur Radio equipment. Interference becomes more severe with higher power or closer spacing to the signal source. The ARRL's Technical Information Service web page provides information on all kinds of RFI and the means to correct it. [*Ham Radio License Manual*, page 5-19]

T7B04 **Which of the following is a way to reduce or eliminate interference by an amateur transmitter to a nearby telephone?**

A. Put a filter on the amateur transmitter
B. Reduce the microphone gain
C. Reduce the SWR on the transmitter transmission line
D. Put a RF filter on the telephone

D The major cause of telephone interference comes from telephones that were not equipped with interference protection when they were manufactured. (Reference: FCC CIB Telephone Interference Bulletin) Radio frequency energy from your amateur transmitter may be strong enough to cause the telephone to act as a receiver through direct detection, the most common form of interference to telephones. As with receiver overload, there is nothing you can do at the transmitter to cure the interference. Interference protection measures must be taken at, or in, the telephone in question. [*Ham Radio License Manual*, page 5-21]

T7B05 How can overload of a non-amateur radio or TV receiver by an amateur signal be reduced or eliminated?

A. Block the amateur signal with a filter at the antenna input of the affected receiver
B. Block the interfering signal with a filter on the amateur transmitter
C. Switch the transmitter from FM to SSB
D. Switch the transmitter to a narrow-band mode

A Very strong signals may overwhelm a receiver's ability to reject them. Receiver overload is a common type of interference to TV and FM-broadcast receivers. Symptoms of overload include severe interference on all channels of a TV or FM receiver or an amateur may hear noise across an entire band when the strong signal is present. If adding attenuation (either by turning on a receiver's attenuator or removing an antenna) causes the interference to disappear, it's probably caused by overload. The strong signal can be blocked or reduced to harmless levels by putting an RF filter at the antenna input of the receiver. [*Ham Radio License Manual*, page 5-21]

T7B06 Which of the following actions should you take if a neighbor tells you that your station's transmissions are interfering with their radio or TV reception?

A. Make sure that your station is functioning properly and that it does not cause interference to your own radio or television when it is tuned to the same channel
B. Immediately turn off your transmitter and contact the nearest FCC office for assistance
C. Tell them that your license gives you the right to transmit and nothing can be done to reduce the interference
D. Install a harmonic doubler on the output of your transmitter and tune it until the interference is eliminated

A You should first make sure that your equipment is operating properly and not creating spurious signals on the TV channels. Check for interference to your own TV and broadcast radios. If you observe interference, stop operating and cure the problem before you go back on the air. Your neighbor may still experience interference but you will be able to demonstrate that your equipment is operating properly. [*Ham Radio License Manual*, page 5-22]

T7B07 Which of the following may be useful in correcting a radio frequency interference problem?

A. Snap-on ferrite chokes
B. Low-pass and high-pass filters
C. Band-reject and band-pass filters
D. All of these choices are correct

D Knowing how to use the different kinds of filters and RF blocking materials such as ferrite cores is an important technique for amateurs. If you are having an interference problem, members of your club may be able to help you. The ARRL also maintains an RF Interference resource web page as part of the Technical Information Service area on the ARRL website. [*Ham Radio License Manual*, page 5-19]

T7B08 What should you do if something in a neighbor's home is causing harmful interference to your amateur station?

A. Work with your neighbor to identify the offending device
B. Politely inform your neighbor about the rules that prohibit the use of devices which cause interference
C. Check your station and make sure it meets the standards of good amateur practice
D. All of these choices are correct

D This is one of the benefits of being a licensed station — protection against interference. While this means that the responsibility for stopping the interference may lie with your neighbor, you should still make sure the interference is not occurring because of some problem in your station. If you are sure your station is operating properly, you may need to politely educate your neighbor about the interference and help identify the offending device. [*Ham Radio License Manual*, page 5-23]

T7B09 What is a Part 15 device?

A. An unlicensed device that may emit low powered radio signals on frequencies used by a licensed service
B. A type of amateur radio that can legally be used in the citizen's band
C. A device for long distance communications using special codes sanctioned by the International Amateur Radio Union
D. A type of test set used to determine whether a transmitter is in compliance with FCC regulation 91.15

A Part 15 refers to the section of the FCC Rules that permits unlicensed devices to use radio frequencies as part of their function. This includes cordless phones, baby monitors, audio and video relay devices, and wireless computer data links. These devices are allowed to use the radio spectrum, but are not permitted to cause interference to stations in licensed services, such as Amateur Radio. Users of Part 15 devices are required to accept interference to the devices from stations in a licensed service, such as Amateur Radio or a broadcast station. These rules must be printed either directly on the device or in the owner's manual supplied with the device. [*Ham Radio License Manual*, page 5-23]

T7B10 What might be the problem if you receive a report that your audio signal through the repeater is distorted or unintelligible?

A. Your transmitter may be slightly off frequency
B. Your batteries may be running low
C. You could be in a bad location
D. All of these choices are correct

D A station that is slightly off-frequency will be strong, but distorted. This sometimes happens when a radio control key gets bumped, changing frequency by a small amount. If accidentally pressing a control key on your radio is a frequent problem (smaller radios are particularly prone to it), try using the LOCK feature of your radio to disable unintended key presses. You could also be causing excessive deviation by speaking too loudly into the microphone. Either lower your voice or hold the microphone farther away from your mouth. Weak or low batteries can also cause distorted audio by causing the transmitter's modulator circuitry to function improperly. [*Ham Radio License Manual*, page 6-12]

T7B11 What is a symptom of RF feedback in a transmitter or transceiver?

A. Excessive SWR at the antenna connection
B. The transmitter will not stay on the desired frequency
C. Reports of garbled, distorted, or unintelligible transmissions
D. Frequent blowing of power supply fuses

C It is not unusual for RF current flowing in sensitive audio cables or data cables to interfere with your station's normal function, just as your strong transmitter signal might be picked up and detected by a neighbor's telephone or audio system. "RF feedback" via a microphone cable can cause distorted transmitted audio, for example. [*Ham Radio License Manual*, page 5-24]

T7B12 What might be the first step to resolve cable TV interference from your ham radio transmission?

A. Add a low pass filter to the TV antenna input
B. Add a high pass filter to the TV antenna input
C. Add a preamplifier to the TV antenna input
D. Be sure all TV coaxial connectors are installed properly

D Properly installed cable TV connections prevent interference from amateur transmissions by keeping ham signals out of the receiver input. However, if the connection is loose or the connector is damaged or improperly installed, external signals such as amateur transmissions can get into the received signal path and cause interference. [*Ham Radio License Manual*, page 5-21]

T7C Antenna measurements and troubleshooting: measuring SWR; dummy loads; coaxial cables; feed line failure modes

T7C01 What is the primary purpose of a dummy load?

A. To prevent the radiation of signals when making tests
B. To prevent over-modulation of your transmitter
C. To improve the radiation from your antenna
D. To improve the signal to noise ratio of your receiver

A To avoid interfering with other stations while you're adjusting your transmitter or measuring its output power, it's a good idea to use a dummy load. A dummy load is a heavy-duty resistor that can absorb and dissipate the output power from a transmitter. Because the dummy load absorbs all of the transmitter output and radiates it as heat instead of radio waves, there is no signal to interfere with other hams. You should always use a dummy load when testing a transmitter unless you absolutely have to make an on-the-air test. [*Ham Radio License Manual*, page 5-4]

T7C02 Which of the following instruments can be used to determine if an antenna is resonant at the desired operating frequency?

A. A VTVM
B. An antenna analyzer
C. A Q meter
D. A frequency counter

B An antenna analyzer consists of a very low-power signal source with an adjustable frequency and one or more meters or displays to show the impedance and SWR without using a transmitter's output signal that is strong enough to cause interference. [*Ham Radio License Manual*, page 4-19]

T7C03 What, in general terms, is standing wave ratio (SWR)?

A. A measure of how well a load is matched to a transmission line
B. The ratio of high to low impedance in a feed line
C. The transmitter efficiency ratio
D. An indication of the quality of your station's ground connection

A Standing wave ratio, or SWR, is caused by a mismatch of the characteristic impedance of the feed line and that of a load (such as an antenna) attached to the feed line. Some of the energy delivered by the feed line is reflected at the load and interferes with the incoming energy, creating stationary patterns called standing waves. The magnitude of the SWR is the ratio of the maximum and minimum values of thc standing wave pattern. [*Ham Radio License Manual*, page 4-10]

T7C04 What reading on an SWR meter indicates a perfect impedance match between the antenna and the feed line?

A. 2 to 1
B. 1 to 3
C. 1 to 1
D. 10 to 1

C When there is no reflected power there are no standing waves and the SWR is 1:1. This condition is called a perfect match. [*Ham Radio License Manual*, page 4-10]

T7C05 What is the approximate SWR value above which the protection circuits in most solid-state transmitters begin to reduce transmitter power?

A. 2 to 1
B. 1 to 2
C. 6 to 1
D. 10 to 1

A One possible effect of high SWR is that the standing waves cause voltages to increase in the feed line. This can occur at the transmitter's output where the feed line is connected. The higher voltages can damage a transmitter's output circuits. Most amateur transmitting equipment is designed to work at full power with an SWR of 2:1 or lower. SWR greater than 2:1 may cause the transmitter to reduce power automatically in order to protect its output circuit. [*Ham Radio License Manual*, page 4-10]

T7C06 What does an SWR reading of 4:1 indicate?

A. Loss of –4 dB
B. Good impedance match
C. Gain of +4 dB
D. Impedance mismatch

D SWR greater than 1:1 is called an impedance mismatch or mismatch. [*Ham Radio License Manual*, page 4-10]

T7C07 What happens to power lost in a feed line?

A. It increases the SWR
B. It comes back into your transmitter and could cause damage
C. It is converted into heat
D. It can cause distortion of your signal

C As power travels through a feed line, some of it is absorbed by the insulation in the line and some is lost in the resistance of the conductors themselves. These losses are in the form of heat, just as if the power was dissipated by a resistor. Feed lines used at radio frequencies use special materials and construction methods to minimize power being dissipated as heat by feed line loss and to avoid signals leaking in or out. [*Ham Radio License Manual*, page 4-8]

T7C08 **What instrument other than an SWR meter could you use to determine if a feed line and antenna are properly matched?**

A. Voltmeter
B. Ohmmeter
C. Iambic pentameter
D. Directional wattmeter

D A directional wattmeter can tell you how much energy is flowing in each direction along a feed line. From the relative amounts of power flowing in each direction, you can calculate SWR. By minimizing reflected power, you can adjust your antenna for the best impedance match. [*Ham Radio License Manual*, page 4-18]

T7C09 **Which of the following is the most common cause for failure of coaxial cables?**

A. Moisture contamination
B. Gamma rays
C. The velocity factor exceeds 1.0
D. Overloading

A Coaxial cables must be protected. The performance of coaxial cable depends on the integrity of its outer coating, the jacket. Nicks, cuts and scrapes can all breach the jacket allowing moisture contamination, the most common cause of coaxial cable failure. Prolonged exposure to the ultraviolet (UV) in sunlight will also cause the plastic in the jacket to degrade, causing small cracks that allow water into the cable. To protect the cable against UV damage the jacket usually contains a pigment that absorbs and blocks the UV. Capillary action of the strands making up the braided outer shield can also draw water into coaxial cable. Once in the cable, the water causes both corrosion and heat losses in the shield. Water can also get into coaxial cable through an improperly sealed connection. [*Ham Radio License Manual*, page 4-16]

T7C10 **Why should the outer jacket of coaxial cable be resistant to ultraviolet light?**

A. Ultraviolet resistant jackets prevent harmonic radiation
B. Ultraviolet light can increase losses in the cable's jacket
C. Ultraviolet and RF signals can mix together, causing interference
D. Ultraviolet light can damage the jacket and allow water to enter the cable

D See question T7C09. [*Ham Radio License Manual*, page 4-16]

T7C11 What is a disadvantage of air core coaxial cable when compared to foam or solid dielectric types?

A. It has more loss per foot
B. It cannot be used for VHF or UHF antennas
C. It requires special techniques to prevent water absorption
D. It cannot be used at below freezing temperatures

C Coax connectors exposed to the weather must be carefully waterproofed. If you use low-loss air-core coax, you will need to pay extra attention to waterproofing the connectors because special techniques are required to prevent water absorption by this cable. [*Ham Radio License Manual*, page 4-17]

T7C12 Which of the following is a common use of coaxial cable?

A. Carrying dc power from a vehicle battery to a mobile radio
B. Carrying RF signals between a radio and antenna
C. Securing masts, tubing, and other cylindrical objects on towers
D. Connecting data signals from a TNC to a computer

B Coaxial cable is specially made to carry RF signals without allowing them to radiate or be affected by other signals. Materials that make up coaxial cable are chosen to minimize the amount of signal power lost as the signal travels through the cable. This makes coaxial cable an ideal choice to carry RF signals between a radio and an antenna. [*Ham Radio License Manual*, page 4-9]

T7C13 What does a dummy load consist of?

A. A high-gain amplifier and a TR switch
B. A non-inductive resistor and a heat sink
C. A low voltage power supply and a DC relay
D. A 50 ohm reactance used to terminate a transmission line

B Dummy loads convert RF power to heat and are used for transmitter testing so that test signals are not radiated on the air. [*Ham Radio License Manual*, page 5-4]

T7D Basic repair and testing: soldering; using basic test instruments; connecting a voltmeter, ammeter, or ohmmeter

T7D01 Which instrument would you use to measure electric potential or electromotive force?

A. An ammeter
B. A voltmeter
C. A wavemeter
D. An ohmmeter

B Electromotive force (EMF) is another term for voltage, which is measured in volts. A meter used to measure voltage is called a voltmeter. [*Ham Radio License Manual*, page 3-1]

T7D02 What is the correct way to connect a voltmeter to a circuit?

A. In series with the circuit
B. In parallel with the circuit
C. In quadrature with the circuit
D. In phase with the circuit

B Voltmeters are connected in parallel with a component (also termed "across a component") or circuit to measure voltage. Voltage is always measured from one point to another or with respect to some reference voltage. [*Ham Radio License Manual*, page 3-3]

T7D03 How is an ammeter usually connected to a circuit?

A. In series with the circuit
B. In parallel with the circuit
C. In quadrature with the circuit
D. In phase with the circuit

A Ammeters are connected in series with a component or circuit to measure current. Some meters calibrated in amperes are really voltmeters that measure the voltage across a small resistor in series with the current. Be sure to determine which type of meter you have before using it. [*Ham Radio License Manual*, page 3-3]

T7D04 Which instrument is used to measure electric current?

A. An ohmmeter
B. A wavemeter
C. A voltmeter
D. An ammeter

D The units of current flow are amperes and a meter used to measure current is called an ammeter. [*Ham Radio License Manual*, page 3-1]

T7D05 What instrument is used to measure resistance?

A. An oscilloscope
B. A spectrum analyzer
C. A noise bridge
D. An ohmmeter

D The units of resistance are ohms and a meter used to measure resistance is called an ohmmeter. [*Ham Radio License Manual*, page 3-4]

T7D06 Which of the following might damage a multimeter?

A. Measuring a voltage too small for the chosen scale
B. Leaving the meter in the milliamps position overnight
C. Attempting to measure voltage when using the resistance setting
D. Not allowing it to warm up properly

C The flexibility afforded by a multimeter also means that it is important to use the meter properly. Trying to measure voltage or connecting the probes to an energized circuit when the meter is set to measure resistance is a common way to damage a multimeter, for example. [*Ham Radio License Manual*, page 3-3]

T7D07 Which of the following measurements are commonly made using a multimeter?

A. SWR and RF power
B. Signal strength and noise
C. Impedance and reactance
D. Voltage and resistance

D The typical multimeter uses a switch and different sets of input connections to select which parameter and range of values to measure. The most common measurements are voltage, resistance, and current. The convenience of having all the different functions in a single instrument more than outweighs the extra complexity of learning to use a multimeter. [*Ham Radio License Manual*, page 3-3]

T7D08 Which of the following types of solder is best for radio and electronic use?

A. Acid-core solder
B. Silver solder
C. Rosin-core solder
D. Aluminum solder

C Learning how to install your own coax connectors and make cables not only saves money but allows you to make repairs at home and under emergency conditions! Start by reading the ARRL Technical Information Service's online article "The Art of Soldering." Follow up with "Connectors For All Occasions, Parts 1 and 2" on the same web page. You'll learn what kind of solder to use for electronics (rosin-core), what a "cold" solder joint looks like (it has a grainy or dull surface), and many other useful tips. [*Ham Radio License Manual*, page 4-17]

T7D09 What is the characteristic appearance of a cold solder joint?

A. Dark black spots
B. A bright or shiny surface
C. A grainy or dull surface
D. A greenish tint

C See question T7D08. [*Ham Radio License Manual*, page 4-17]

T7D10 What is probably happening when an ohmmeter, connected across an unpowered circuit, initially indicates a low resistance and then shows increasing resistance with time?

A. The ohmmeter is defective
B. The circuit contains a large capacitor
C. The circuit contains a large inductor
D. The circuit is a relaxation oscillator

B This is an indication that a large (high-value) capacitor is part of the circuit. As the meter transfers energy to the capacitor it begins to appear more like a high-value resistor to the multimeter's dc resistance measuring circuit. You'll find out more about capacitor charging when you begin studying for your General class license. [*Ham Radio License Manual*, page 3-3]

T7D11 **Which of the following precautions should be taken when measuring circuit resistance with an ohmmeter?**

A. Ensure that the applied voltages are correct
B. Ensure that the circuit is not powered
C. Ensure that the circuit is grounded
D. Ensure that the circuit is operating at the correct frequency

B A circuit with power applied can cause enough current to flow in a multimeter's sensitive resistance measurement circuits to cause measurement errors or even damage the meter. Always make sure power is removed from a circuit before making resistance measurements. [*Ham Radio License Manual*, page 3-3]

T7D12 **Which of the following precautions should be taken when measuring high voltages with a voltmeter?**

A. Ensure that the voltmeter has very low impedance
B. Ensure that the voltmeter and leads are rated for use at the voltages to be measured
C. Ensure that the circuit is grounded through the voltmeter
D. Ensure that the voltmeter is set to the correct frequency

B Even though a meter's enclosure and leads are insulated, connecting them to voltages beyond their rating can result in "flashovers" from the live circuit to the user, causing electrocution or severe shock. Remember that the peak voltage of an ac waveform is approximately 40 percent higher than its RMS value and the peak-to-peak value is nearly 3 times the RMS value! [*Ham Radio License Manual*, page 3-3]

Operating Modes and Special Operations

SUBELEMENT T8 — Modulation modes: amateur satellite operation; operating activities; non-voice communications [4 Exam Questions — 4 Groups]

T8A Modulation modes: bandwidth of various signals; choice of emission type

T8A01 **Which of the following is a form of amplitude modulation?**

A. Spread spectrum
B. Packet radio
C. Single sideband
D. Phase shift keying

C An amplitude modulation (AM) signal consists of a carrier and two sidebands, one higher in frequency than the carrier (the upper sideband, USB) and one lower (the lower sideband, LSB). A single-sideband (SSB) signal is created by removing the carrier and one of the sidebands. [*Ham Radio License Manual*, page 2-9]

T8A02 **What type of modulation is most commonly used for VHF packet radio transmissions?**

A. FM
B. SSB
C. AM
D. Spread spectrum

A FM can be used for data signals, such as those for packet radio on VHF and UHF. The data is sent as audio tones by using the FM transmitter voice input and speaker output. This allows inexpensive FM voice radios to be used. [*Ham Radio License Manual*, page 2-10]

T8A03 Which type of voice modulation is most often used for long-distance (weak signal) contacts on the VHF and UHF bands?

A. FM
B. DRM
C. SSB
D. PM

C Because the SSB signal's power is concentrated into a narrow bandwidth, it is possible to communicate with SSB over much longer ranges and in poorer conditions than with FM or AM, particularly on the VHF and UHF bands. That is why the VHF and UHF DXers and contest operators use SSB signals. [*Ham Radio License Manual*, page 2-11]

T8A04 Which type of modulation is most commonly used for VHF and UHF voice repeaters?

A. AM
B. SSB
C. PSK
D. FM

D Frequency modulation (FM) and its close relative phase modulation (PM) are used in repeater communication because of the mode's superior rejection of noise and static. [*Ham Radio License Manual*, page 2-10]

T8A05 Which of the following types of emission has the narrowest bandwidth?

A. FM voice
B. SSB voice
C. CW
D. Slow-scan TV

C CW signals consist of a continuous wave signal on a single frequency that is turned on and off in the patterns that make up the Morse code. As a result, CW has the narrowest bandwidth of any mode copied directly by the human ear. [*Ham Radio License Manual*, page 2-10]

T8A06 **Which sideband is normally used for 10 meter HF, VHF and UHF single-sideband communications?**

A. Upper sideband
B. Lower sideband
C. Suppressed sideband
D. Inverted sideband

A There is no technical reason for choosing USB over LSB. However, in order to make communications easier, amateur radio has standardized on the following conventions:

• Below 10 MHz, LSB is used

• Above 10 MHz, USB is used — including all of the VHF and UHF bands

This convention is even programmed into radio equipment as the normal operating mode! There is one exception: amateurs are required to use USB for voice communication on the five 60 meter band (5 MHz) channels. [*Ham Radio License Manual*, page 2-11]

T8A07 **What is the primary advantage of single sideband over FM for voice transmissions?**

A. SSB signals are easier to tune
B. SSB signals are less susceptible to interference
C. SSB signals have narrower bandwidth
D. All of these choices are correct

C Whereas an FM signal may have a bandwidth of 5 to 15 kHz, SSB signals only require 2 to 3 kHz of spectrum. This increases the usable range of SSB signals compared to FM signals of equal power. (See question T8A03.) [*Ham Radio License Manual*, page 2-11]

T8A08 **What is the approximate bandwidth of a single sideband voice signal?**

A. 1 kHz
B. 3 kHz
C. 6 kHz
D. 15 kHz

B See the following table of Signal Bandwidths for various modes. [*Ham Radio License Manual*, page 2-5]

Signal Bandwidths

Type of Signal	*Typical Bandwidth*
AM amateur voice	6 kHz
AM broadcast	10 kHz
Commercial video broadcast	6 MHz (see note)
SSB voice	2 to 3 kHz
SSB digital	500 to 3000 Hz (0.5 to 3 kHz)
CW	100 to 300 Hz (0.1 to 0.3 kHz)
FM amateur voice	10 to 15 kHz
FM broadcast	150 kHz

Note: Within each 6 MHz channel, there may be from four to five digitally-compressed audio-video programs, each with a 1.2-1.5 MHz bandwidth. Amateurs will continue to use the older analog NTSC format for fast-scan television for the foreseeable future and the bandwidth of those signals is approximately 6 MHz.

T8A09 **What is the approximate bandwidth of a VHF repeater FM phone signal?**

A. Less than 500 Hz
B. About 150 kHz
C. Between 10 and 15 kHz
D. Between 50 and 125 kHz

C See the table of Signal Bandwidths. [*Ham Radio License Manual*, page 2-5]

T8A10 **What is the typical bandwidth of analog fast-scan TV transmissions on the 70 cm band?**

A. More than 10 MHz
B. About 6 MHz
C. About 3 MHz
D. About 1 MHz

B See the table of Signal Bandwidths. [*Ham Radio License Manual*, page 2-5]

T8A11 **What is the approximate maximum bandwidth required to transmit a CW signal?**

A. 2.4 kHz
B. 150 Hz
C. 1000 Hz
D. 15 kHz

B See the table of Signal Bandwidths. [*Ham Radio License Manual*, page 2-5]

T8B Amateur satellite operation; Doppler shift, basic orbits, operating protocols; control operator, transmitter power considerations; satellite tracking; digital modes

T8B01 **Who may be the control operator of a station communicating through an amateur satellite or space station?**

A. Only an Amateur Extra Class operator
B. A General Class licensee or higher licensee who has a satellite operator certification
C. Only an Amateur Extra Class operator who is also an AMSAT member
D. Any amateur whose license privileges allow them to transmit on the satellite uplink frequency

D [97.301, 97.207(c)] — The control operator of a station communicating through an amateur satellite may hold any class of amateur license that has privileges to transmit in the satellite's input frequency range, regardless of the frequency of the satellite's downlink signal. [*Ham Radio License Manual*, page 6-30]

T8B02 **How much transmitter power should be used on the uplink frequency of an amateur satellite or space station?**

A. The maximum power of your transmitter
B. The minimum amount of power needed to complete the contact
C. No more than half the rating of your linear amplifier
D. Never more than 1 watt

B [97.313] Using minimum power is particularly important when using satellites because of their very limited power sources. By transmitting with excessive power, you can actually cause the signals of other stations using the satellite to be reduced as they are relayed by the satellite. [*Ham Radio License Manual*, page 6-31]

T8B03 **Which of the following are provided by satellite tracking programs?**

A. Maps showing the real-time position of the satellite track over the earth
B. The time, azimuth, and elevation of the start, maximum altitude, and end of a pass
C. The apparent frequency of the satellite transmission, including effects of Doppler shift
D. All of these answers are correct

D In order to communicate through a satellite, you need to know all of these things. Using the published Keplerian elements (data about the satellite's orbit) the software can make all of the necessary calculations. Check the AMSAT website (**www.amsat.org**) for information on available software. [*Ham Radio License Manual*, page 6-31]

T8B04 **Which amateur stations may make contact with an amateur station on the International Space Station using 2 meter and 70 cm band amateur radio frequencies?**

A. Only members of amateur radio clubs at NASA facilities
B. Any amateur holding a Technician or higher class license
C. Only the astronaut's family members who are hams
D. You cannot talk to the ISS on amateur radio frequencies

B [97.301, 97.207(c)] The astronaut-amateurs on board the ISS operate according to the same FCC Rules that earthbound hams do, so there is no reason that communications between them should be restricted based on license class. Any amateur with a license that permits communications on the VHF frequencies used by the ISS is welcome to make contact! [*Ham Radio License Manual*, page 6-30]

T8B05 **What is a satellite beacon?**

A. The primary transmit antenna on the satellite
B. An indicator light that shows where to point your antenna
C. A reflective surface on the satellite
D. A transmission from a space station that contains information about a satellite

D Like its terrestrial counterparts, a satellite's beacon transmits communications so that propagation to the satellite and information about the satellite can be observed. Receiving the beacon from a satellite indicates that it is within range and contacts with it or through it can be attempted. [*Ham Radio License Manual*, page 6-30]

T8B06 **Which of the following are inputs to a satellite tracking program?**

A. The weight of the satellite
B. The Keplerian elements
C. The last observed time of zero Doppler shift
D. All of these answers are correct

B See question T8B03. [*Ham Radio License Manual*, page 6-31]

T8B07 **With regard to satellite communications, what is Doppler shift?**

A. A change in the satellite orbit
B. A mode where the satellite receives signals on one band and transmits on another
C. An observed change in signal frequency caused by the relative motion between the satellite and the earth station
D. A special digital communications mode for some satellites

C Doppler shift or Doppler effect describes the way the downlink frequency of a satellite varies by several kilohertz during an orbit due to its motion relative to the receiving station. As the satellite is moving toward the receiving station, the frequency of the satellite's signal will increase by a small amount. After the satellite passes overhead, the frequency of the signal will begin to decrease. This is the same effect that causes the sound of a vehicle's horn or whistle to change pitch as it moves past you. [*Ham Radio License Manual*, page 6-30]

T8B08 **What is meant by the statement that a satellite is operating in mode U/V?**

A. The satellite uplink is in the 15 meter band and the downlink is in the 10 meter band
B. The satellite uplink is in the 70 cm band and the downlink is in the 2 meter band
C. The satellite operates using ultraviolet frequencies
D. The satellite frequencies are usually variable

B The satellite's operational mode specifies the bands on which it is transmitting and receiving. Most satellites only have one mode, but some have several that can be controlled by ground stations. Mode is specified as two letters separated by a slash. The first letter indicates the uplink band and the second letter indicates the downlink band. For example, the uplink for a satellite in U/V mode is in the UHF band (70 cm) and a downlink in the VHF band (2 meters). [*Ham Radio License Manual*, page 6-31]

T8B09 What causes spin fading when referring to satellite signals?

A. Circular polarized noise interference radiated from the sun
B. Rotation of the satellite and its antennas
C. Doppler shift of the received signal
D. Interfering signals within the satellite uplink band

B Most satellites spin to stabilize their orientation in space. Spin fading is caused by rotation of the satellite and its antennas with respect to the receiving station. As the satellite spins, its signal will strengthen and weaken depending on the orientation of the antennas. [*Ham Radio License Manual*, page 6-31]

T8B10 What do the initials LEO tell you about an amateur satellite?

A. The satellite battery is in Low Energy Operation mode
B. The satellite is performing a Lunar Ejection Orbit maneuver
C. The satellite is in a Low Earth Orbit
D. The satellite uses Light Emitting Optics

C Low Earth Orbit is the term for satellites orbiting up to about 1600 miles above the Earth, including the International Space Station. LEO satellites have nearly circular orbits and are only visible from Earth for a few minutes. Higher orbits are referred to as Medium Earth Orbit. Farthest out are the Geosynchronous Earth Orbit satellites, whose orbital period matches exactly the rotation of the Earth so that the satellite appears to stay above one spot on Earth. [*Ham Radio License Manual*, page 6-30]

T8B11 What is a commonly used method of sending signals to and from a digital satellite?

A. USB AFSK
B. PSK31
C. FM Packet
D. WSJT

C Some satellites are equipped with FM packet radio stations. These satellites can operate as digipeaters and sometimes as bulletin board stations. Many CubeSat and FunCube satellites use packet radio to transmit their telemetry and other experimental data to earth stations. [*Ham Radio License Manual*, page 6-31]

T8C Operating activities: radio direction finding; radio control; contests; linking over the Internet; grid locators

T8C01 Which of the following methods is used to locate sources of noise interference or jamming?

A. Echolocation
B. Doppler radar
C. Radio direction finding
D. Phase locking

C There are a number of techniques that allow a radio operator to determine the bearing to a radio transmitter. Using them is called direction finding. Exercises in which hams try to find a hidden transmitter are also called foxhunts or bunny hunts. In recent years a new type of outdoor radiosport has reached US shores from Europe and Asia — radio direction finding or RDF. Held as organized events, direction finding is a hybrid of the radio fox hunt using orienteering skills to navigate outdoors with map and compass. The US Amateur Radio Direction Finding organization (**www.ardfusa.com**) is just one of a number of national groups in this worldwide sport, especially popular with teens and young adults. If you are a hiker or camper, then you might be interested in applying your outdoor skills to ARDF. [*Ham Radio License Manual*, page 6-29]

T8C02 Which of these items would be useful for a hidden transmitter hunt?

A. Calibrated SWR meter
B. A directional antenna
C. A calibrated noise bridge
D. All of these choices are correct

B You don't need much in the way of equipment. A portable radio with a signal strength indicator and a handheld or portable directional antenna, such as a small Yagi beam, will work just fine. The point of the exercise is to determine direction to a radio source, the hidden transmitter, and to do so requires a directional antenna. [*Ham Radio License Manual*, page 6-29]

T8C03 What popular operating activity involves contacting as many stations as possible during a specified period of time?

A. Contesting
B. Net operations
C. Public service events
D. Simulated emergency exercises

A A radio contest or radiosport event consists of making as many contacts as you can with stations in a targeted area or on specific bands. Contests last for only a few hours or maybe all weekend. They're a great way to improve your station capabilities and operating skills. See the ARRL Contest Branch website for an event calendar and links to contest rules. [*Ham Radio License Manual*, page 6-28]

T8C04 Which of the following is good procedure when contacting another station in a radio contest?

A. Be sure to sign only the last two letters of your call if there is a pileup calling the station
B. Work the station twice to be sure that you are in his log
C. Send only the minimum information needed for proper identification and the contest exchange
D. All of these choices are correct

C Check the contest rules and listen to a few contacts during the contest to find out what information is required in the contest exchange. [*Ham Radio License Manual*, page 6-28]

T8C05 What is a grid locator?

A. A letter-number designator assigned to a geographic location
B. A letter-number designator assigned to an azimuth and elevation
C. An instrument for neutralizing a final amplifier
D. An instrument for radio direction finding

A The Maidenhead Locator System (named after the city in England where it was developed) divides the Earth's surface into grids organized by latitude and longitude. The designator for each grid square consists of two letter and two numbers, such as CN87 or FM13. For example, ARRL Headquarters in Newington, Connecticut is located in grid square FN31. A further two letters can be added for greater precision, such as FN31pq for the precise location of the ARRL station. (Learn more about grid squares at the ARRL website.) [*Ham Radio License Manual*, page 6-4]

T8C06 How is access to an IRLP node accomplished?

A. By obtaining a password which is sent via voice to the node
B. By using DTMF signals
C. By entering the proper Internet password
D. By using CTCSS tone codes

B IRLP stands for the Internet Relay Linking Project by which audio can be sent from one repeater to another using the Internet. Choosing and activating a link between repeaters is done by sending DTMF (Touch-Tone) tones to the local repeater, which then routes digitized audio to the designated repeater. [*Ham Radio License Manual*, page 6-19]

T8C07 What is the maximum power allowed when transmitting telecommand signals to radio controlled models?

A. 500 milliwatts
B. 1 watt
C. 25 watts
D. 1500 watts

B [97.215(c)] — Amateurs may transmit telecommand signals with an output power of up to 1 watt. Although the signals do not identify the licensee on the air, radio control (RC) modelers are required to display their name, call sign and address on the RC transmitter. [*Ham Radio License Manual*, page 6-33]

T8C08 What is required in place of on-air station identification when sending signals to a radio control model using amateur frequencies?

A. Voice identification must be transmitted every 10 minutes
B. Morse code ID must be sent once per hour
C. A label indicating the licensee's name, call sign and address must be affixed to the transmitter
D. A flag must be affixed to the transmitter antenna with the station call sign in 1 inch high letters or larger

C [97.215(a)] — See question T8C07. [*Ham Radio License Manual*, page 6-33]

T8C09 How might you obtain a list of active nodes that use VoIP?

A. From the FCC Rulebook
B. From your local emergency coordinator
C. From a repeater directory
D. From the local repeater frequency coordinator

C Both the IRLP and EchoLink systems maintain an Internet directory of repeaters and relays (both referred to as nodes) participating in their systems. You can browse the directories to find nodes (repeaters and Internet voice server systems) currently on the air. [*Ham Radio License Manual*, page 6-19]

T8C10 How do you select a specific IRLP node when using a portable transceiver?

A. Choose a specific CTCSS tone
B. Choose the correct DSC tone
C. Access the repeater autopatch
D. Use the keypad to transmit the IRLP node ID

D To initiate an IRLP or Echolink contact, the initiating station must know the repeater control code to request an IRLP connection — this is the ON code. The ON code varies from repeater to repeater and obtaining it may require membership in a club. Once the ON code is entered via your keypad, the four-digit code for the IRLP node — a destination repeater — is entered. The code then directs the repeater controller to establish the Internet link to the desired node. [*Ham Radio License Manual*, page 6-19]

T8C11 What name is given to an amateur radio station that is used to connect other amateur stations to the Internet?

A. A gateway
B. A repeater
C. A digipeater
D. A beacon

A An Internet gateway is a special kind of digital station that provides a connection to the Internet for data transmitted by Amateur Radio from other stations. Most gateways are set up to forward messages. The most common examples are packet radio bulletin board systems (BBS) and the Winlink RMS stations. Messages with a recognized Internet email address can be sent and retrieved over these systems. Another type of gateway provides direct Internet connectivity so that a computer running standard web browser software can connect to any Internet address. [*Ham Radio License Manual*, page 5-14]

T8C12 What is meant by Voice Over Internet Protocol (VoIP) as used in amateur radio?

A. A set of rules specifying how to identify your station when linked over the Internet to another station
B. A set of guidelines for working DX during contests using Internet access
C. A technique for measuring the modulation quality of a transmitter using remote sites monitored via the Internet
D. A method of delivering voice communications over the Internet using digital techniques

D VoIP is a formal digital communication protocol used for transferring audio and voice signals over the Internet. Amateurs employ VoIP primarily to transfer signals between repeaters. [*Ham Radio License Manual*, page 6-19]

T8C13 What is the Internet Radio Linking Project (IRLP)?

A. A technique to connect amateur radio systems, such as repeaters, via the Internet using Voice Over Internet Protocol
B. A system for providing access to websites via amateur radio
C. A system for informing amateurs in real time of the frequency of active DX stations
D. A technique for measuring signal strength of an amateur transmitter via the Internet

A See question T8C06. [*Ham Radio License Manual*, page 6-18]

T8D Non-voice communications: image signals; digital modes; CW; packet; PSK31; APRS; error detection and correction; NTSC

T8D01 Which of the following is an example of a digital communications method?

A. Packet
B. PSK31
C. MFSK
D. All of these choices are correct

D Packet radio uses a version of the computer-to-computer X.25 data exchange protocol that was adapted to suit amateur use becoming the AX.25 protocol. PSK31 is an abbreviation for phase-shift keying at 31 baud for low-rate keyboard-to-keyboard communications in a narrow bandwidth. MFSK uses frequency-shift keying of multiple carriers for data communications. Packet radio is generally used on VHF and UHF frequencies while PSK31 and MFSK are primarily used on the HF bands. [*Ham Radio License Manual*, page 5-9]

T8D02 What does the term APRS mean?

A. Automatic Packet Reporting System
B. Associated Public Radio Station
C. Auto Planning Radio Set-up
D. Advanced Polar Radio System

A The Automatic Packet Reporting System (APRS) was created by WB4APR as a way of transmitting GPS location information over Amateur Radio using a system of digipeaters. These relay points forward the position information and call sign to a system of server computers via the Internet. Once the information is stored on the servers websites can access the data and show the position of the station on maps in various ways. See Figure T8-1. [*Ham Radio License Manual*, page 5-11]

T8D03 Which of the following devices provides data to the transmitter when sending automatic position reports from a mobile amateur radio station?

A. The vehicle speedometer
B. A WWV receiver
C. A connection to a broadcast FM sub-carrier receiver
D. A Global Positioning System receiver

D See question T8D02. [*Ham Radio License Manual*, page 5-11]

T8D04 What type of transmission is indicated by the term NTSC?

A. A Normal Transmission mode in Static Circuit
B. A special mode for earth satellite uplink
C. An analog fast scan color TV signal
D. A frame compression scheme for TV signals

C NTSC stands for the National Television System Committee, which developed the standards that define broadcast fast-scan analog television signals. (For many years, broadcast TV stations in the United States used the NTSC standard for their analog TV broadcasts before they converted to a digital TV signal standard in 2009.) Amateurs adopted the NTSC standard for amateur television (ATV) transmissions. RS-170 is the standard that describes fast-scan video signals before they are sent over the air. These signals can be received on a commercial analog TV receiver equipped with a suitable frequency converter to tune amateur frequencies. The *ARRL Handbook, ARRL Operating Manual* and Technical Information Service pages on the ARRL website provide more information on ATV. [*Ham Radio License Manual*, page 6-32]

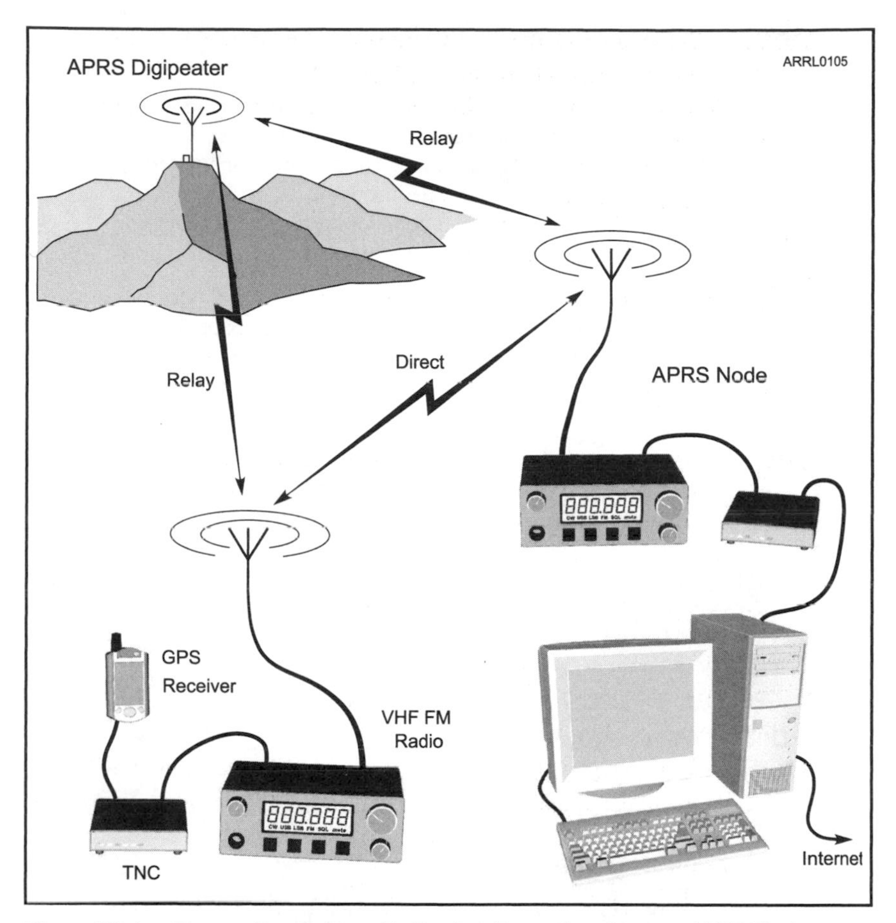

Figure T8-1 — To use the Automatic Packet Reporting System (APRS), the output of a GPS receiver is connected to a packet radio TNC and a VHF radio. Position information and the call sign of the reporting station are then transferred to the ARPS system by an APRS node, either directly or via a digipeater. Station location can then be viewed via the Internet.

T8D05 Which of the following is an application of APRS (Automatic Packet Reporting System)?

A. Providing real time tactical digital communications in conjunction with a map showing the locations of stations
B. Showing automatically the number of packets transmitted via PACTOR during a specific time interval
C. Providing voice over Internet connection between repeaters
D. Providing information on the number of stations signed into a repeater

A See question T8D02. [*Ham Radio License Manual*, page 5-11]

T8D06 What does the abbreviation PSK mean?

A. Pulse Shift Keying
B. Phase Shift Keying
C. Packet Short Keying
D. Phased Slide Keying

B In phase-shift keying (PSK), the phase of a signal is varied in order to convey information. Amateurs use PSK of an audio signal to transmit information. The audio signal can be transmitted as an AM or FM signal. [*Ham Radio License Manual*, page 5-11]

T8D07 What is PSK31?

A. A high-rate data transmission mode
B. A method of reducing noise interference to FM signals
C. A method of compressing digital television signals
D. A low-rate data transmission mode

D See question T8D01. [*Ham Radio License Manual*, page 5-11]

T8D08 Which of the following may be included in packet transmissions?

A. A check sum which permits error detection
B. A header which contains the call sign of the station to which the information is being sent
C. Automatic repeat request in case of error
D. All of these choices are correct

D Packet radio sends data in bursts called packets. Each packet consists of a header and data. The header contains information about the packet and the call sign of the destination station. The header also includes a checksum that allows the receiver to detect errors. If an error is detected, the receiver automatically requests that the packet be retransmitted until the data is received properly. This is called ARQ for automatic repeat request. [*Ham Radio License Manual*, page 5-10]

T8D09 What code is used when sending CW in the amateur bands?

A. Baudot
B. Hamming
C. International Morse
D. Gray

C Amateurs use the International Morse code for CW transmissions. [*Ham Radio License Manual*, page 5-9]

T8D10 Which of the following can be used to transmit CW in the amateur bands?

A. Straight Key
B. Electronic Keyer
C. Computer Keyboard
D. All of these choices are correct

D For sending Morse code, a key is used to turn the transmitter output signal on and off. Morse code's dots and dashes are known as the elements of the code. When using a straight key, the operator generates the dots and dashes manually. This is called hand keying. The J-38 is a common type of straight key and is popular with beginners. Once you are skilled at "the code," you'll want to go faster by using a keyer. This electronic device turns contact closures from a Morse paddle into a stream of Morse code elements. A keyer may be a standalone device or it can be built in to a transceiver. A paddle is a pair of levers mounted side by side, each having its own set of contacts, one for dots and one for dashes. Keyers and paddles can generate Morse code much faster than by using a straight key. Computer software can also send CW directly from the keyboard by using a keying interface connected to the key input of a radio. [*Ham Radio License Manual*, page 5-6]

T8D11 What is an ARQ transmission system?

A. A special transmission format limited to video signals
B. A system used to encrypt command signals to an amateur radio satellite
C. A digital scheme whereby the receiving station detects errors and sends a request to the sending station to retransmit the information
D. A method of compressing the data in a message so more information can be sent in a shorter time

C ARQ stands for Automatic Repeat request or Automatic Repeat Query. This is a type of digital communications protocol in which the receiving station requests a retransmission of data if it detects an error. PACTOR and packet radio are ARQ protocols. [*Ham Radio License Manual*, page 5-10]

Subelement T9

Antennas and Feed Lines

SUBELEMENT T9 — Antennas and feed lines [2 Exam Questions — 2 Groups]

T9A Antennas: vertical and horizontal polarization; concept of gain; common portable and mobile antennas; relationships between antenna length and frequency

T9A01 **What is a beam antenna?**

A. An antenna built from aluminum I-beams
B. An omnidirectional antenna invented by Clarence Beam
C. An antenna that concentrates signals in one direction
D. An antenna that reverses the phase of received signals

C The term "beam" is used in the same sense as a flashlight beam, meaning that the antenna is directional just as the reflector of a flashlight acts to concentrate the light bulb's "signal" in one direction. The beam doesn't create more power, only focuses it. Beams can be used to increase signal level at a distant station or to reject interference or noise. [*Ham Radio License Manual*, page 4-14]

T9A02 **Which of the following is true regarding vertical antennas?**

A. The magnetic field is perpendicular to the Earth
B. The electric field is perpendicular to the Earth
C. The phase is inverted
D. The phase is reversed

B The vertical antenna has that name for two reasons. First, the antenna element or elements are vertically oriented! Second, the vertical orientation of an antenna element means the electric field component of the radio waves it radiates is vertical or perpendicular to the surface of the Earth. The antenna and its radiated waves are thus vertically polarized. [*Ham Radio License Manual*, page 4-6]

T9A03 Which of the following describes a simple dipole mounted so the conductor is parallel to the Earth's surface?

A. A ground wave antenna
B. A horizontally polarized antenna
C. A rhombic antenna
D. A vertically polarized antenna

B Like the vertical, a dipole is called a horizontal antenna for two reasons. The antenna element is oriented horizontally so that the radiated radio wave is horizontally polarized — its electric field is oriented parallel to the Earth's surface. [*Ham Radio License Manual*, page 4-11]

T9A04 What is a disadvantage of the "rubber duck" antenna supplied with most handheld radio transceivers?

A. It does not transmit or receive as effectively as a full-sized antenna
B. It transmits a circularly polarized signal
C. If the rubber end cap is lost it will unravel very quickly
D. All of these choices are correct

A When you buy a new VHF or UHF hand-held transceiver, it will usually have a flexible rubber-coated antenna commonly called a "rubber duck." This antenna is inexpensive, small, lightweight and difficult to break. On the other hand, its performance is not nearly as good as a full-sized antenna, such as a mobile antenna or a telescoping whip. [*Ham Radio License Manual*, page 4-13]

T9A05 How would you change a dipole antenna to make it resonant on a higher frequency?

A. Lengthen it
B. Insert coils in series with radiating wires
C. Shorten it
D. Add capacitive loading to the ends of the radiating wires

C Remember that wavelength and frequency change in opposite directions; as one increases, the other decreases and vice versa. Thus, to make the dipole resonant on a higher frequency, shorten it. All three of the other choices will lower the dipole's resonant frequency. [*Ham Radio License Manual*, page 4-12]

T9A06 What type of antennas are the quad, Yagi, and dish?

A. Non-resonant antennas
B. Loop antennas
C. Directional antennas
D. Isotropic antennas

C All three of these antennas focus the radiated energy toward one direction. The quad and Yagi work by having groups of elements work together to reinforce radiated energy in the desired direction. The dish antenna works like a flashlight, reflecting energy so that a great deal of it is focused in one direction. [*Ham Radio License Manual*, page 4-15]

T9A07 What is a good reason not to use a "rubber duck" antenna inside your car?

A. Signals can be significantly weaker than when it is outside of the vehicle
B. It might cause your radio to overheat
C. The SWR might decrease, decreasing the signal strength
D. All of these choices are correct

A The vehicle's metal roof and doors act like shields, trapping the radio waves inside. Some of the signal gets out through the windows (unless they're tinted by a thin metal coating), but it's much weaker than if radiated by an external antenna. [*Ham Radio License Manual*, page 4-13]

T9A08 What is the approximate length, in inches, of a quarter-wavelength vertical antenna for 146 MHz?

A. 112
B. 50
C. 19
D. 12

C Start with the formula for estimating the length of a 1/2-wavelength dipole: Length (in feet) = 468 / frequency (in MHz) = 468 / 146 = 3.2 feet. Convert to inches by multiplying by 12 = 38.5. Because this is a quarter-wavelength antenna, divide the result by two to get about 19 inches. [*Ham Radio License Manual*, page 4-11]

T9A09 What is the approximate length, in inches, of a 6 meter ½-wavelength wire dipole antenna?

A. 6
B. 50
C. 112
D. 236

C Since you already know that the wavelength is 6 meters, ½-wavelength is 3 meters. Convert meters to feet by multiplying by 3.1 = 9.3 feet. Convert to inches by multiplying by 12 = 112 inches. [*Ham Radio License Manual*, page 4-11]

T9A10 In which direction is the radiation strongest from a half-wave dipole antenna in free space?

A. Equally in all directions
B. Off the ends of the antenna
C. Broadside to the antenna
D. In the direction of the feed line

C A dipole radiates strongest broadside to the axis of the dipole and weakest off the ends. The radiation pattern for a dipole isolated in space looks like a donut or bagel as seen in Figure T9-1, which shows both two- and three-dimensional patterns. The two-dimensional pattern shows what the three-dimensional pattern would look like if cut through the axis of the dipole. [*Ham Radio License Manual*, page 4-11]

T9A11 What is meant by the gain of an antenna?

A. The additional power that is added to the transmitter power
B. The additional power that is lost in the antenna when transmitting on a higher frequency
C. The increase in signal strength in a specified direction when compared to a reference antenna
D. The increase in impedance on receive or transmit compared to a reference antenna

C The concentration of radio signals in a specific direction is called *gain*. (Antenna gain should not be confused with the gain of a transistor.) Antenna gain is a measure of how much signal strength is increased in a specified direction when compared to the signal from a reference antenna in the same direction. Gain aids communication in the preferred direction by increasing transmitted and received signal strengths. [*Ham Radio License Manual*, page 4-6]

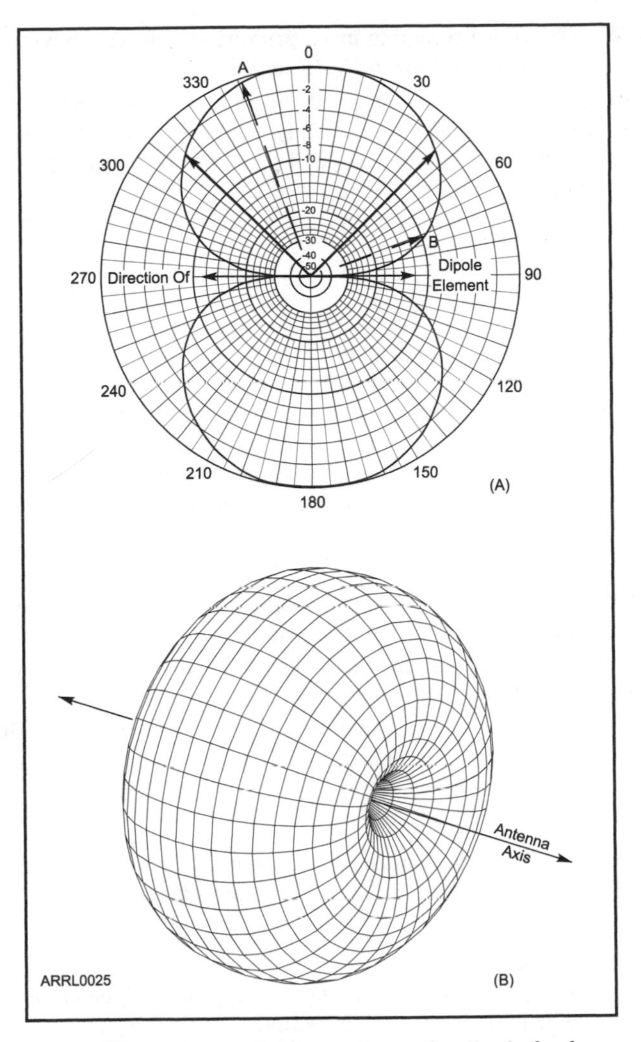

Figure T9-1 — The radiation pattern of a dipole far from ground (in free-space). At (A) the pattern is shown in a plane containing the dipole. The lengths of the arrows indicate the relative strength of the radiated power in that direction. The dipole radiates best broadside to its length. At (B) the 3-D pattern shows radiated strength in all directions.

T9A12 What is a reason to use a properly mounted 5⁄8 wavelength antenna for VHF or UHF mobile service?

A. It offers a lower angle of radiation and more gain than a ¼ wavelength antenna and usually provides improved coverage
B. It features a very high angle of radiation and is better for communicating via a repeater
C. The ⅝ wavelength antenna completely eliminates distortion caused by reflected signals
D. The ⅝ wavelength antenna offers a 10-times power gain over a ¼ wavelength design

A The peak angle for the radiation pattern of a ⅝-wavelength ground-plane antenna mounted vertically and in the clear is a bit closer to horizontal than for a ¼-wavelength ground-plane. This lower angle focuses transmission and reception closer to the horizon, which usually increases communication range. In areas where the repeater is high above average terrain, such as in mountainous areas, the lower peak angle of radiation may not be helpful. [*Ham Radio License Manual*, page 4-13]

T9A13 Why are VHF or UHF mobile antennas often mounted in the center of the vehicle roof?

A. Roof mounts have the lowest possible SWR of any mounting configuration
B. Only roof mounting can guarantee a vertically polarized signal
C. A roof mounted antenna normally provides the most uniform radiation pattern
D. Roof mounted antennas are always the easiest to install

C Mounting the vertical antenna in the middle of the roof will generally provide the best signal strength at the receiving station, no matter how the vehicle is oriented. [*Ham Radio License Manual*, page 4-13]

T9A14 Which of the following terms describes a type of loading when referring to an antenna?

A. Inserting an inductor in the radiating portion of the antenna to make it electrically longer
B. Inserting a resistor in the radiating portion of the antenna to make it resonant
C. Installing a spring at the base of the antenna to absorb the effects of collisions with other objects
D. Making the antenna heavier so it will resist wind effects when in motion

A Loading an antenna refers to adding inductance or capacitance to lower the frequency at which the antenna is resonant. This allows antennas to present a resonant feed point impedance on frequencies below their natural resonant frequency, in effect lengthening them electrically. [*Ham Radio License Manual*, page 4-13]

T9B Feed lines: types of feed lines; attenuation vs. frequency; SWR concepts; matching; weather protection; choosing RF connectors and feed lines

T9B01 Why is it important to have a low SWR in an antenna system that uses coaxial cable feed line?

A. To reduce television interference
B. To allow the efficient transfer of power and reduce losses
C. To prolong antenna life
D. All of these choices are correct

B Power reflected from a mismatched antenna bounces back and forth in the feed line. Some is transferred to the antenna on each trip but with each pass through the feed line, some power is lost as heat. Low SWR reduces losses in the feed line because less power is reflected from the antenna. As SWR increases, more power is reflected and more power is lost. Since coaxial feed line has higher losses than open-wire line, low SWR is more important if coaxial feed line is used. [*Ham Radio License Manual*, page 4-10]

T9B02 What is the impedance of the most commonly used coaxial cable in typical amateur radio installations?

A. 8 ohms
B. 50 ohms
C. 600 ohms
D. 12 ohms

B Most coaxial cable used by amateurs has a characteristic impedance (Z_0) of 50 Ω. Coaxial cables used for video and cable television have a Z_0 of 75 Ω. Open-wire or twin-lead feed lines have a Z_0 of 300 to 450 Ω. [*Ham Radio License Manual*, page 4-9]

T9B03 Why is coaxial cable used more often than any other feed line for amateur radio antenna systems?

A. It is easy to use and requires few special installation considerations
B. It has less loss than any other type of feed line
C. It can handle more power than any other type of feed line
D. It is less expensive than any other types of feed line

A Coaxial cable or coax is easy to work with because it is a single, compact cable. All of the energy being conducted by the feed line is completely contained within the cable, so the cable can be run alongside or even within metallic trays, conduits and towers. It can be coiled up and placed next to other cables without effect. While it may have more loss than open-wire line and weigh a bit more per foot of length, coaxial cable's ease of use makes it the most practical choice in many installations. [*Ham Radio License Manual*, page 4-9]

T9B04 What does an antenna tuner do?

A. It matches the antenna system impedance to the transceiver's output impedance
B. It helps a receiver automatically tune in weak stations
C. It allows an antenna to be used on both transmit and receive
D. It automatically selects the proper antenna for the frequency band being used

A If the SWR at the end of the feed line is too high for the radio to operate properly, devices called antenna tuners (or impedance matchers or transmatches) or are connected at the output of the transmitter as shown in Figure T9-2. An antenna tuner is adjusted until the SWR measured at the transmitter output is acceptably close to 1:1. This means the antenna system's impedance has been matched to that of the transmitter output. An antenna tuner doesn't really tune the antenna, it just adjusts the impedance at the input to the antenna system. Think of the antenna tuner as an electrical gearbox that lets the engine (the transmitter) run at the speed it likes no matter how fast the tires are turning (feed point impedance). [*Ham Radio License Manual*, page 4-18]

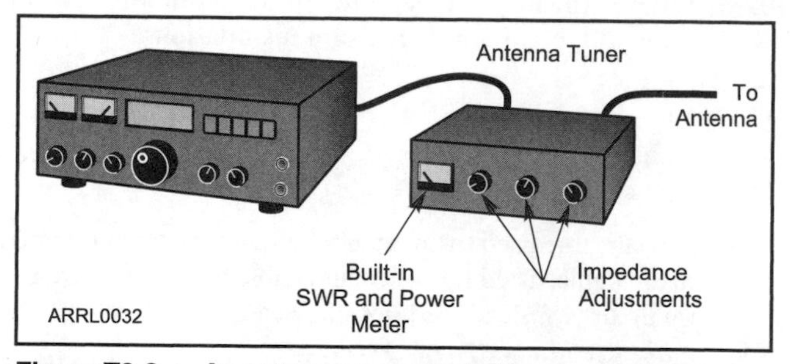

Figure T9-2 — An antenna tuner acts like an electrical version of a mechanical gearbox. By adjusting the tuner's controls, the impedance present at the end of the feed line can be converted to the impedance that best suits the transceiver's output circuits, usually 50 Ω.

T9B05 What generally happens as the frequency of a signal passing through coaxial cable is increased?

A. The apparent SWR increases
B. The reflected power increases
C. The characteristic impedance increases
D. The loss increases

D Feed lines used at radio frequencies use special materials and construction methods to minimize power being dissipated as heat (called feed line loss) and to avoid signals leaking in or out. Feed line loss increases with frequency for all types of coaxial feed lines. [*Ham Radio License Manual*, page 4-8]

T9B06 Which of the following connectors is most suitable for frequencies above 400 MHz?

A. A UHF (PL-259/SO-239) connector
B. A Type N connector
C. An RS-213 connector
D. A DB-25 connector

B Which connector to use depends on the frequency of the signals being used. The UHF series of connectors are the most widely-used for HF equipment. (UHF does not stand for "ultra-high frequency" in this case.) Above 400 MHz, the Type N connectors are preferred. You'll find both UHF and N connectors on 6, 2 and 1.25 meter equipment. [*Ham Radio License Manual*, page 4-17]

T9B07 Which of the following is true of PL-259 type coax connectors?

A. They are preferred for microwave operation
B. They are water tight
C. They are commonly used at HF frequencies
D. They are a bayonet type connector

C See question T9B06. [*Ham Radio License Manual*, page 4-17]

T9B08 Why should coax connectors exposed to the weather be sealed against water intrusion?

A. To prevent an increase in feed line loss
B. To prevent interference to telephones
C. To keep the jacket from becoming loose
D. All of these choices are correct

A Coax connectors exposed to the weather must be carefully waterproofed because water in coaxial cable degrades the effectiveness of the braided shield and dramatically increases losses. (See also question T7C09.) [*Ham Radio License Manual*, page 4-17]

T9B09 What might cause erratic changes in SWR readings?

A. The transmitter is being modulated
B. A loose connection in an antenna or a feed line
C. The transmitter is being over-modulated
D. Interference from other stations is distorting your signal

B Remember that SWR is caused by a mismatch between feed line impedance and load or antenna impedance. If there is a loose connection where the load is attached (at the antenna feed point), then the erratic connection acts like an erratically changing load impedance. The result is an erratic change in SWR. [*Ham Radio License Manual*, page 4-10]

T9B10 **What electrical difference exists between the smaller RG-58 and larger RG-8 coaxial cables?**

A. There is no significant difference between the two types
B. RG-58 cable has less loss at a given frequency
C. RG-8 cable has less loss at a given frequency
D. RG-58 cable can handle higher power levels

C An important characteristic of coax is feed line loss. Loss is specified in dB per 100 feet of cable at a certain frequency. The following table gives cable loss at 30 MHz (close to the 10 meter band) and at 150 MHz (close to the 2 meter band). In general, if made from the same materials, a larger diameter cable such as RG-8 will have less loss than a small cable such as RG-58. [*Ham Radio License Manual*, page 4-16]

Cable Type	*Impedance (Ω)*	*Loss per 100 feet (in dB) at 30 MHz*	*Loss per 100 feet (in dB) at 150 MHz*
RG-8	50	1.1	2.5
RG-58	50	2.5	5.6

T9B11 **Which of the following types of feed line has the lowest loss at VHF and UHF?**

A. 50-ohm flexible coax
B. Multi-conductor unbalanced cable
C. Air-insulated hard line
D. 75-ohm flexible coax

C A special type of air-insulated coaxial feed line is called *hard line* because its shield is made from a semi-flexible solid tube of aluminum or copper. This limits how much the cable can be bent or flexed, but hard line has the lowest loss of any type of coaxial feed line. [*Ham Radio License Manual*, page 4-9]

Electrical and RF Safety

SUBELEMENT T0 — Electrical safety: AC and DC power circuits; antenna installation; RF hazards [3 Exam Questions — 3 Groups]

T0A Power circuits and hazards: hazardous voltages; fuses and circuit breakers; grounding; lightning protection; battery safety; electrical code compliance

T0A01 **Which of the following is a safety hazard of a 12-volt storage battery?**

A. Touching both terminals with the hands can cause electrical shock
B. Shorting the terminals can cause burns, fire, or an explosion
C. RF emissions from the battery
D. All of these choices are correct

B Storage batteries, such as those used for vehicle starter and power service, are made to deliver large amounts of energy very quickly. If they are short-circuited with a tool, wire, or sheet metal, the resulting currents are very high — high enough to melt metal! The resulting heat can cause severe burns, start a fire, or even cause the battery to explode. Take extreme care around batteries and never perform electrical work on a vehicle without disconnecting the battery first! [*Ham Radio License Manual*, page 9-3]

T0A02 **How does current flowing through the body cause a health hazard?**

A. By heating tissue
B. It disrupts the electrical functions of cells
C. It causes involuntary muscle contractions
D. All of these choices are correct

D Burns caused by dc current or low-frequency ac current are a result of resistance to current in the skin, either through it to the body's interior or along it from point to point. The current creates heat and that's what results in the burn. Electrical current through the body can disrupt the electrical function of cells. Currents of more than a few mA can also cause involuntary muscle contractions, portrayed as jerking and jumping on TV and in the movies. No joking matter, muscle spasms can cause falls and sudden large movements. The sudden pulling back of an outstretched hand or finger that comes in contact with an energized conductor is the result of arm muscles contracting involuntarily. [*Ham Radio License Manual*, page 9-2]

T0A03 **What is connected to the green wire in a three-wire electrical AC plug?**

A. Neutral
B. Hot
C. Safety ground
D. The white wire

C State and national electrical-safety codes require the three-wire power cords on many 120-V tools and appliances. Power supplies and station equipment use similar connections. Two of the conductors (the "hot" and the "neutral" wires) power the device. The third conductor (the safety ground wire) connects to the metal frame or chassis of the device. The "hot" wire is usually black or red. The "neutral" wire is white. The frame/ground wire is green or bare. See Figure T0-1. [*Ham Radio License Manual*, page 9-4]

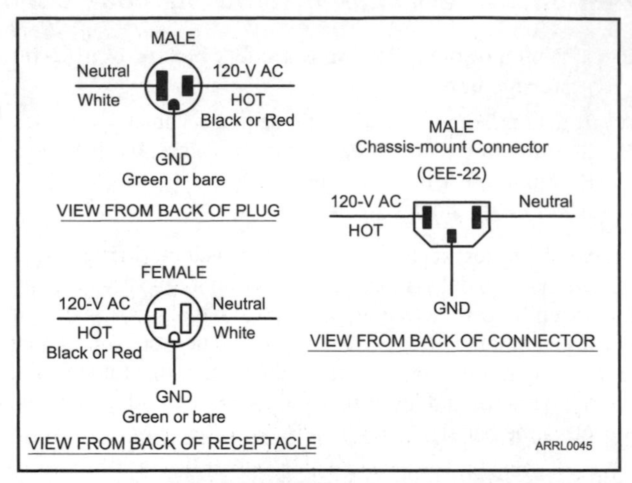

Figure T0-1 — The correct wiring technique for 120 V ac power cords and receptacles. The white wire is neutral and the green wire is the safety ground. The hot wire can be either black or red. These receptacles are shown from the back, or wiring side.

T0A04 **What is the purpose of a fuse in an electrical circuit?**

A. To prevent power supply ripple from damaging a circuit
B. To interrupt power in case of overload
C. To limit current to prevent shocks
D. All of these choices are correct

B A fuse consists of a thin strip of metal that melts at relatively low temperatures. Current flow causes the metal to heat up and when too much current flows, the metal melts, breaking the circuit and interrupting power. [*Ham Radio License Manual*, page 3-12]

T0A05 Why is it unwise to install a 20-ampere fuse in the place of a 5-ampere fuse?

A. The larger fuse would be likely to blow because it is rated for higher current
B. The power supply ripple would greatly increase
C. Excessive current could cause a fire
D. All of these choices are correct

C Never replace a fuse or circuit breaker with one rated for a larger current, since that allows more current to flow in response to a fault in the equipment. The higher current could overheat wires and cause a fire. Determine what problem caused the fuse to blow and make repairs so that higher current does not make the damage worse or destroy the equipment entirely. [*Ham Radio License Manual*, page 3-12]

T0A06 What is a good way to guard against electrical shock at your station?

A. Use three-wire cords and plugs for all AC powered equipment
B. Connect all AC powered station equipment to a common safety ground
C. Use a circuit protected by a ground-fault interrupter
D. All of these choices are correct

D The best way to protect against electrical shock is to make sure that all exposed parts of your station equipment are at the same potential or voltage. That potential should, of course, be ground potential. You'll want to make sure that you have a common safety ground, meaning that all grounds should be connected together. Three-wire cords and plugs make sure metal equipment enclosures are connected to the safety ground. A ground-fault interrupter detects current imbalance between the hot and neutral wires, indicating an undesired current path that creates a shock hazard, and turns off power to the circuit. [*Ham Radio License Manual*, page 9-3]

T0A07 Which of these precautions should be taken when installing devices for lightning protection in a coaxial cable feed line?

A. Include a parallel bypass switch for each protector so that it can be switched out of the circuit when running high power
B. Include a series switch in the ground line of each protector to prevent RF overload from inadvertently damaging the protector
C. Keep the ground wires from each protector separate and connected to station ground
D. Ground all of the protectors to a common plate which is in turn connected to an external ground

D Even though amateur antennas and towers are struck no more frequently than tall trees or other nearby structures, it is wise to take some precautionary steps. This is especially true for stations in areas with frequent severe weather and lightning. Lightning protection is intended to provide fire protection for your home since most of the damage to a home resulting from a lightning strike is from fire. Starting at your antennas, all towers, masts, and antenna mounts should be grounded according to your local building codes. This is done at the base, or in the case of roof mounts, though a large-diameter wire to a ground rod. Ground connections should be as short and direct as possible — avoid sharp bends. Where cables enter the house, use lightning arrestors grounded to a common plate that is in turn connected to a nearby external ground such as a ground rod. [*Ham Radio License Manual*, page 9-5]

T0A08 What safety equipment should always be included in home-built equipment that is powered from 120V AC power circuits?

A. A fuse or circuit breaker in series with the AC hot conductor
B. An AC voltmeter across the incoming power source
C. An inductor in series with the AC power source
D. A capacitor across the AC power source

A If you decide to replace or run new wiring for your station, either have a licensed electrician do the wiring or inspect your work. Be sure to follow the wiring standard: hot is black (or occasionally red); neutral is white; and ground is green or bare. Use cable and wire sufficiently rated for the expected current load. Use the proper size fuses and circuit breakers. If you build your own equipment and power it from the ac lines, be sure to always install a fuse or circuit breaker in series with the ac hot conductor. [*Ham Radio License Manual*, page 9-4]

T0A09 What kind of hazard is presented by a conventional 12-volt storage battery?

A. It emits ozone which can be harmful to the atmosphere
B. Shock hazard due to high voltage
C. Explosive gas can collect if not properly vented
D. All of these choices are correct

C Storage batteries pack a lot of energy into a small volume, but with that energy comes the need to treat them carefully. The liquid acid in the battery is extremely corrosive and will eat holes in anything organic (including your clothes and skin!) During charging, hydrogen gas is given off by the battery and can be explosive if there is no ventilation to disperse it. (See also question T0A01.) [*Ham Radio License Manual*, page 5-18]

T0A10 What can happen if a lead-acid storage battery is charged or discharged too quickly?

A. The battery could overheat and give off flammable gas or explode
B. The voltage can become reversed
C. The memory effect will reduce the capacity of the battery
D. All of these choices are correct

A Trying to get energy into or out of any battery too quickly can cause it to overheat. Excessive charging can cause hydrogen gas to build up faster than the battery can vent it to the outside air. Overheating is particularly dangerous with storage batteries, since they hold so much energy. Use a battery charger that is designed to work with storage batteries and keep the load on the battery to within the manufacturer's specifications. [*Ham Radio License Manual*, page 5-18]

T0A11 What kind of hazard might exist in a power supply when it is turned off and disconnected?

A. Static electricity could damage the grounding system
B. Circulating currents inside the transformer might cause damage
C. The fuse might blow if you remove the cover
D. You might receive an electric shock from the charge stored in large capacitors

D A capacitor is designed to store charge, so it should not be a surprise that a capacitor can remain charged even after power is removed. It is good practice to place bleeder resistors across large capacitors to slowly allow the charge to dissipate. Never assume that capacitors, particularly those in high-voltage circuits, are discharged. Measure them with a voltmeter first or use a grounding stick to discharge them. [*Ham Radio License Manual*, page 9-3]

T0B Antenna safety: tower safety; erecting an antenna support; overhead power lines; installing an antenna

T0B01 When should members of a tower work team wear a hard hat and safety glasses?

A. At all times except when climbing the tower
B. At all times except when belted firmly to the tower
C. At all times when any work is being done on the tower
D. Only when the tower exceeds 30 feet in height

C A piece of hardware or a tool will be traveling 40 mph by the time it falls 60 feet! Ouch! Whenever the crew is working on the tower, wear your protective gear — even if you are not near the base of the tower. If a falling object hits the tower or a guy wire, it can bounce a long way. Try to stay out from under the crew at work at the top. And while you're at it, remember the sun block! [*Ham Radio License Manual*, page 9-13]

T0B02 What is a good precaution to observe before climbing an antenna tower?

A. Make sure that you wear a grounded wrist strap
B. Remove all tower grounding connections
C. Put on a climbing harness and safety glasses
D. All of the these choices are correct

C Make sure your climbing harness and glasses and hard hat are in good condition — then use them! Climb slowly, with a lanyard around the tower. It's not a race, so take your time. Once at the top, work slowly, thinking out each move before you make it. Have a backup plan and never work on a tower without a ground crew or someone to keep an eye on you. [*Ham Radio License Manual*, page 9-13]

T0B03 Under what circumstances is it safe to climb a tower without a helper or observer?

A. When no electrical work is being performed
B. When no mechanical work is being performed
C. When the work being done is not more than 20 feet above the ground
D. Never

D Having a ground crew is important; avoid climbing alone whenever possible because it's never safe. If you do climb alone, take along a handheld radio so that you can call for help if needed. [*Ham Radio License Manual*, page 9-14]

T0B04 Which of the following is an important safety precaution to observe when putting up an antenna tower?

A. Wear a ground strap connected to your wrist at all times
B. Insulate the base of the tower to avoid lightning strikes
C. Look for and stay clear of any overhead electrical wires
D. All of these choices are correct

C If power lines ever come into contact with your antenna, you could be electrocuted. The only safe place to install an antenna tower is in a location that is well clear of any power lines. Before you put up a tower, look for any overhead electrical wires. Make sure that the tower is installed where there is no possibility of contact between the lines and the tower if the guy wires should ever break or the tower fall. [*Ham Radio License Manual*, page 9-12]

T0B05 What is the purpose of a gin pole?

A. To temporarily replace guy wires
B. To be used in place of a safety harness
C. To lift tower sections or antennas
D. To provide a temporary ground

C A gin pole is a temporary mast used to lift materials such as antennas or tower sections so that you do not have to hoist things directly. This is much safer than a direct lift and supports the materials while you work on them. [*Ham Radio License Manual*, page 9-14]

T0B06 What is the minimum safe distance from a power line to allow when installing an antenna?

A. Half the width of your property
B. The height of the power line above ground
C. 1/2 wavelength at the operating frequency
D. So that if the antenna falls unexpectedly, no part of it can come closer than 10 feet to the power wires

D Ten feet of separation is the minimum amount for safety if the tower and antenna fall directly toward the power lines. Figure the separation from the very top of the antenna or with the antenna oriented so that it is closest to the power lines. Allow for more separation whenever possible. [*Ham Radio License Manual*, page 9-12]

T0B07 Which of the following is an important safety rule to remember when using a crank-up tower?

A. This type of tower must never be painted
B. This type of tower must never be grounded
C. This type of tower must never be climbed unless it is in the fully retracted position
D. All of these choices are correct

C Climbing a crank-up tower places your hands and feet between rungs and braces that can do a lot of damage if the tower sections slip or a cable breaks. Not only should you be sure the tower is fully nested, but also place a safety block such as piece of pipe or a 2×4 between the rungs to prevent movement. [*Ham Radio License Manual*, page 9-14]

T0B08 What is considered to be a proper grounding method for a tower?

A. A single four-foot ground rod, driven into the ground no more than 12 inches from the base
B. A ferrite-core RF choke connected between the tower and ground
C. Separate eight-foot long ground rods for each tower leg, bonded to the tower and each other
D. A connection between the tower base and a cold water pipe

C A tower's safety ground is intended to conduct lightning energy to the earth, reducing the amount traveling along your feed lines. Grounding each leg of the tower balances lightning currents. It also dissipates the charge into the ground as widely as possible. [*Ham Radio License Manual*, page 9-13]

T0B09 Why should you avoid attaching an antenna to a utility pole?

A. The antenna will not work properly because of induced voltages
B. The utility company will charge you an extra monthly fee
C. The antenna could contact high-voltage power wires
D. All of these choices are correct

C Never attach an antenna or guy wire to a utility pole, since a mechanical failure could result in contact with high-voltage lines. [*Ham Radio License Manual*, page 9-12]

T0B10 **Which of the following is true concerning grounding conductors used for lightning protection?**

A. Only non-insulated wire must be used
B. Wires must be carefully routed with precise right-angle bends
C. Sharp bends must be avoided
D. Common grounds must be avoided

C See question T0A07. [*Ham Radio License Manual*, page 9-4]

T0B11 **Which of the following establishes grounding requirements for an amateur radio tower or antenna?**

A. FCC Part 97 Rules
B. Local electrical codes
C. FAA tower lighting regulations
D. Underwriters Laboratories' recommended practices

B See question T0A07. [*Ham Radio License Manual*, page 9-4]

T0B12 **Which of the following is good practice when installing ground wires on a tower for lightning protection?**

A. Put a loop in the ground connection to prevent water damage to the ground system
B. Make sure that all bends in the ground wires are clean, right angle bends
C. Ensure that connections are short and direct
D. All of these choices are correct

C See question T0A07. [*Ham Radio License Manual*, page 9-4]

T0C RF hazards: radiation exposure; proximity to antennas; recognized safe power levels; exposure to others; radiation types; duty cycle

T0C01 **What type of radiation are VHF and UHF radio signals?**

A. Gamma radiation
B. Ionizing radiation
C. Alpha radiation
D. Non-ionizing radiation

D Radio and lower frequency waves are classified as *non-ionizing radiation* because the frequency is too low for them to ionize atoms, no matter how intense the power density of the wave may be. The frequency of ionizing radiation must be higher than that of visible light — ultraviolet, X-rays, gamma rays. Those types of radiation can separate electrons from atoms, creating ions and affecting biochemical processes. [*Ham Radio License Manual*, page 9-5]

T0C02 **Which of the following frequencies has the lowest Maximum Permissible Exposure limit?**

A. 3.5 MHz
B. 50 MHz
C. 440 MHz
D. 1296 MHz

B Frequencies at which the body has the highest SAR are from 30 to 1500 MHz. These are the regions on the MPE graph in Figure T0-2 where the limits for exposure are the lowest. For example, when comparing MPE for amateur bands at 3.5, 50, 440 and 1296 MHz, you can see that MPE is lowest at 50 MHz and highest at 3.5 MHz. [*Ham Radio License Manual*, page 9-7]

Maximum Permissible Exposure (MPE) Limits

Controlled Exposure (6-Minute Average)

Frequency Range (MHz)	*Power Density (mW/cm²)*
0.3-3.0	(100)*
3.0-30	($900/f^2$)*
30-300	1.0
300-1500	f/300
1500-100,000	5

Uncontrolled Exposure (30-Minute Average)

Frequency Range (MHz)	*Magnetic Field Power Density (mW/cm²)*
0.3-1.34	(100)*
1.34-30	($180/f^2$)*
30-300	0.2
300-1500	f/1500
1500-100,000	1.0

f = frequency in MHz
* = Plane-wave equivalent power density

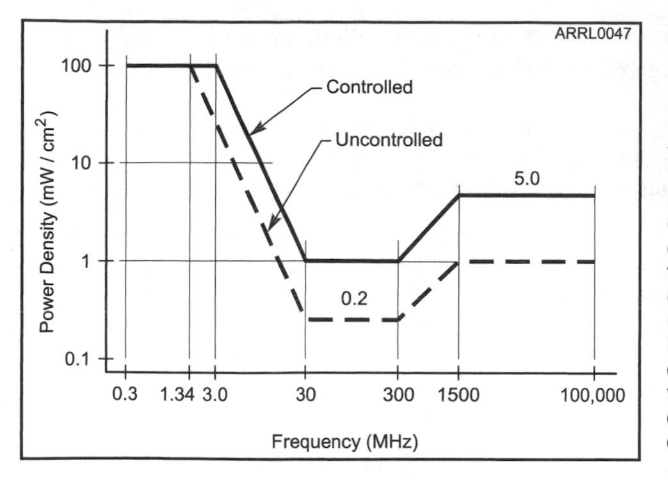

Figure T0-2 — Maximum Permissible Exposure limits vary with frequency because the body responds differently to energy at different frequencies. The controlled and uncontrolled limits refer to the environment in which people are exposed to the RF energy.

T0C03 What is the maximum power level that an amateur radio station may use at VHF frequencies before an RF exposure evaluation is required?

A. 1500 watts PEP transmitter output
B. 1 watt forward power
C. 50 watts PEP at the antenna
D. 50 watts PEP reflected power

C [97.13(c)(1)] — At power levels below 50 watts, the FCC has determined that there is little risk to people. Stations operating with less than 50 watts above 30 MHz are categorically excluded from having to perform a station evaluation. [*Ham Radio License Manual*, page 9-9]

T0C04 What factors affect the RF exposure of people near an amateur station antenna?

A. Frequency and power level of the RF field
B. Distance from the antenna to a person
C. Radiation pattern of the antenna
D. All of these choices are correct

D The human body absorbs less RF energy at some frequencies and more at others. If you decrease your transmitter output power, you decrease the RF energy radiated from your antenna. Placing antennas farther from people reduces the power density to which they are exposed. Finally, the radiation pattern of the antenna affects where RF exposure will be greatest. [*Ham Radio License Manual*, page 9-9]

T0C05 Why do exposure limits vary with frequency?

A. Lower frequency RF fields have more energy than higher frequency fields
B. Lower frequency RF fields do not penetrate the human body
C. Higher frequency RF fields are transient in nature
D. The human body absorbs more RF energy at some frequencies than at others

D At frequencies near the body's natural resonant frequency, RF energy is absorbed more efficiently and maximum heating occurs. In adults, this frequency usually is about 35 MHz if the person is grounded, and about 70 MHz if the person's body is insulated from the ground. Also, body parts may be resonant as well; the adult head, for example is resonant around 400 MHz. Body size thus determines the frequency at which most RF energy is absorbed. As the frequency is increased above resonance, less RF heating generally occurs. [*Ham Radio License Manual*, page 9-6]

T0C06 Which of the following is an acceptable method to determine that your station complies with FCC RF exposure regulations?

A. By calculation based on FCC OET Bulletin 65
B. By calculation based on computer modeling
C. By measurement of field strength using calibrated equipment
D. All of these choices are correct

D [97.13(c)(1)] — You may use a variety of methods to determine that your station complies with FCC RF-exposure regulations. All of the choices given above are correct and valid methods of making that determination. [*Ham Radio License Manual*, page 9-9]

T0C07 What could happen if a person accidentally touched your antenna while you were transmitting?

A. Touching the antenna could cause television interference
B. They might receive a painful RF burn
C. They might develop radiation poisoning
D. All of these choices are correct

B An RF burn is caused by localized heating of the body at the point of contact with the antenna. While painful, it is rarely serious. This is a good reason why you should install your antenna where people cannot accidentally come in contact with it. [*Ham Radio License Manual*, page 9-6]

T0C08 Which of the following actions might amateur operators take to prevent exposure to RF radiation in excess of FCC-supplied limits?

A. Relocate antennas
B. Relocate the transmitter
C. Increase the duty cycle
D. All of these choices are correct

A Anything you can do to reduce power density in an area of concern will reduce RF exposure. This includes relocating your antennas farther from the area, changing the antenna's radiation pattern, and reducing power. The exposure limits also change with frequency, so changing frequency to operate on a band with a higher safe exposure limit is also acceptable. Figure T0-3 shows some options. [*Ham Radio License Manual*, page 9-10]

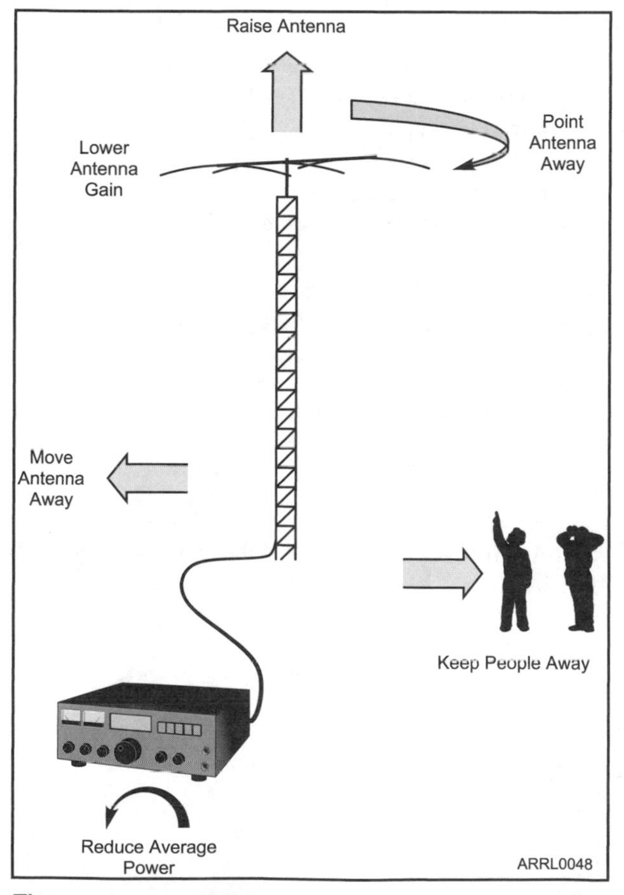

Figure T0-3 — There are many ways to reduce RF exposure to nearby people. Whatever lowers the power density in areas where people are will work. Raising the antenna will even benefit your signal strength to other stations as it lowers power density on the ground!

T0C09 How can you make sure your station stays in compliance with RF safety regulations?

A. By informing the FCC of any changes made in your station
B. By re-evaluating the station whenever an item of equipment is changed
C. By making sure your antennas have low SWR
D. All of these choices are correct

B Whenever you make a change to something that affects power density around your antennas, you should re-evaluate the station. For example, adding an amplifier or changing to an antenna that has more gain will increase power density and you should re-evaluate. If your station is already in compliance and you take a step that decreases RF exposure, such as raising your antennas farther from areas where people are, you don't need to re-evaluate. [*Ham Radio License Manual*, page 9-9]

T0C10 Why is duty cycle one of the factors used to determine safe RF radiation exposure levels?

A. It affects the average exposure of people to radiation
B. It affects the peak exposure of people to radiation
C. It takes into account the antenna feed line loss
D. It takes into account the thermal effects of the final amplifier

A Because the effect of RF exposure is heating, average exposure is what is important. Average exposure during any given period depends on how long the transmitter is operating — which is measured by duty cycle. An emission with a lower duty cycle produces less RF exposure for the same PEP output. Duty cycle is the ratio of the transmitted signal's on-air time to total operating time during the measurement period and has a maximum of 100%. (Duty factor is the same as duty cycle expressed as a fraction, instead of percent, such as 0.25 instead of 25%.) [*Ham Radio License Manual*, page 9-7]

T0C11 What is the definition of duty cycle during the averaging time for RF exposure?

A. The difference between the lowest power output and the highest power output of a transmitter
B. The difference between the PEP and average power output of a transmitter
C. The percentage of time that a transmitter is transmitting
D. The percentage of time that a transmitter is not transmitting

C See question T0C10. [*Ham Radio License Manual,* page 9-7]

T0C12 **How does RF radiation differ from ionizing radiation (radioactivity)?**

A. RF radiation does not have sufficient energy to cause genetic damage
B. RF radiation can only be detected with an RF dosimeter
C. RF radiation is limited in range to a few feet
D. RF radiation is perfectly safe

A See question T0C01. [*Ham Radio License Manual,* page 9-5]

T0C13 **If the averaging time for exposure is 6 minutes, how much power density is permitted if the signal is present for 3 minutes and absent for 3 minutes rather than being present for the entire 6 minutes?**

A. 3 times as much
B. 1/2 as much
C. 2 times as much
D. There is no adjustment allowed for shorter exposure times

C The permitted power density goes up as duty cycle goes down. If the duty cycle is reduced by a factor of 2 (the difference between the transmitter being on for 3 minutes on and off for 3 minutes and the transmitter being on for the full 6 minutes) then the permitted power density is increased by the same amount. (See also question T0C10.) [*Ham Radio License Manual,* page 9-7]

The ARRL Tech Q&A

Index of Advertisers

Quality-built to last...
Mosley... a better antenna!
For a Catalog Call 800-325-4016
www.mosley-electronics.com
POCKET REF
By Thomas J. Glover
POCKET
REF
Thomas J. Glover
ARRL
All the information you'll ever need is right at your fingertips!
This handy pocket-sized guide features tables, charts, drawings, lists, and formulas especially useful for radio amateurs, contractors, students, travelers, electronics hobbyists, craftspeople, and engineers and technicians in virtually every field. Embossed with ARRL's logo and logotype —making this a particularly special edition.
ARRL Order No. 1148
Only $12.95*
*plus shipping and handling
ARRL The national association for AMATEUR RADIO
SHOP DIRECT or call for a dealer near you.
ONLINE WWW.ARRL.ORG/SHOP
ORDER TOLL-FREE 888/277-5289 (US)
HRLM 3/2014
The NEW EZ HANG Square Shot Kit
www.ezhang.com
Now safer and easier to use, you will hit your mark every time, with less chance of misfires hitting the yoke.
Hang your next Wire Antenna the EZ Hang way!
$99.95 + $9.05 for shipping when paying by check
540-286-0176
www.ezhang.com
EZ HANG
32 Princess Gillian Ct.
Fredericksburg, VA 22406
EasyLog5
Since 1989!
Get more information, download a trial version, and order at http://www.easylog.com/northamerica/index.html
or send an email to w5rya@easylog.com
or jbsims_services@att.net.
Check the sales page for discount prices.
www.easylog.com
#1 Ham Radio Books and CDs from ARRL!
www.arrl.org/shop
GIFTS4HAMS.COM
Laser Crafted Gift Items:
License Frames
QSL Storage Chests
Coasters - Mugs - Badges
Paperweights - Key Rings
Pocket Knives - Pens
Rubber Stamps - ID Labels
Ornaments & more ...
visit www.Gifts4Hams.com
HAMPLAQUES.COM
CNC Router & Laser Crafted on Hardwood Solids
License Frames
QSL Storage Chests
CallSign Plaques
Shack Plaques
Club/Organizational Plaques
Plaques for over 20 clubs:
10-10 • CDG • DMC • EARN • EPC • Feld Hell
FISTS • INDEXA • NAQCC • OOTC • OMARC
OMISS • QCWA • QRP-ARCI • SKCC • TCG
Usi • 30MDG • Ozone ARC • LCC • MCARA
JCARA • JARA & more
ASK ABOUT YOUR CLUB
visit www.HamPlaques.com
TUBEDISPLAYS.COM
Display Stands for
Classic, Vintage and Collector Vacuum Tubes
visit www.TubeDisplays.com
ARRL's Visa Signature® Card
Apply Today!
www.arrl.org/visa
Coaxial Dynamics
A CDI INDUSTRIES, INC. COMPANY
SPECIALISTS IN RF TEST EQUIPMENT & COMPONENTS
RF Wattmeters and Plug-in Elements
6800 Lake Abram Drive, Middleburg Hts., Ohio 44130, USA
440-243-1100 • Toll Free: 1-800-COAXIAL • Fax: 440-243-1101
E-Mail: sales@coaxial.com • Web Site: www.coaxial.com

12 STORE BUYING POWER
HAM RADIO OUTLET
WORLDWIDE DISTRIBUTION
D-STAR EXPERTS!
World's LARGEST HAM RADIO INVENTORY in stock for quick delivery
DISCOVER THE POWER OF DSP WITH ICOM!
IC-9100 The All-Round Transceiver
• HF/50MHz 144/430 (440) MHz and 1200MHz*¹ coverage • 100W on HF/50/144MHz, 75W on 430 (440) MHz, 10W on 1200MHz*¹ • Double superheterodyne with image rejection mixer
IC-7800 All Mode Transceiver
• 160-6M @ 200W • Four 32 bit IF-DSPs+ 24 bit AD/DA converters • Two completely independent receivers • +40dBm 3rd order intercept point
IC-7700 Transceiver. The Contester's Rig
• HF + 6m operation • +40dBm ultra high intercept point • IF DSP, user defined filters • 200W output power full duty cycle • Digital voice recorder
IC-7600 All Mode Transceiver
• 100W HF/6m Transceiver, gen cov. receiver • Dual DSP 32 bit • Three roofing filters- 3, 6, 15khz • 5.8 in WQVGA TFT display • Hi-res real time spectrum scope
IC-7410 HF/50MHz Transceiver
• 32-bit floating point DSP unit • Double Conversion Super-Het Receiver • Built-in 15kHz 1st IF Filter • Built-in Band Scope • Large, multi-function LCD • RTTY Demodulator & Decoder • USB for PC control
IC-7200 HF Transceiver
• 160-10M • 100W • Simple & tough with IF DSP • AGC Loop Management • Digital IF Filter • Digital Twin PBT • Digital Noise Reduction • Digital Noise Blanker • USB Port for PC Control
DSP INSTALLED
IC-718 HF Transceiver
• 160-10M* @ 100W • 12V operation • Simple to use • CW Keyer Built-in • One touch band switching • Direct frequency input • VOX Built-in • Band stacking register • IF shift • 101 memories
IC-7100 All Mode Transceiver
• HF/50/144/430/440 MHz Multi-band, Multi-mode, IF DSP • D-STAR DV Mode (Digital Voice + Data) • Intuitive Touch Screen Interface • Built-in RTTY Functions
ID-5100A VHF/UHF Dual BandDigital Txcvr
• Analog FM/D-Star DV Mode • SD Card Slot for Voice & Data Storage • 50W Output on VHF/UHF Bands • Integrated GPS Receiver • AM Airband Dualwatch • FM Analog/DV Repeater List Function
IC-V8000 2M Mobile Transceiver
• 75 watts • Dynamic Memory Scan • CTCSS/DCS encode/decode w/tone scan • Weather alert • Weather channel scan • 200 alphanumeric memories
IC-2300H VHF FM Transceiver
• 65W RF Output Power • 4.5W Audio Output • MIL-STD 810 G Specifications • 207 alphanumeric Memory Channels • Built-in CTCSS/DTCS Encode/Decode • DMS
ID-880H Analog + Digital Dual Bander D-STAR
• D-STAR DV mode operation • DR (D-STAR repeater) mode • Free software download • GPS A mode for easy D-PRS operation • One touch reply button (DV mode) • Wideband receiver
D-STAR ready
IC-92AD Analog + Digital Dual Bander
• 2M/70CM @ 5W • Wide-band RX 495 kHz - 999.9 MHz** • 1304 alphanumeric memories • Dualwatch capability • IPX7 Submersible*** • Optional GPS speaker Mic HM-175GPS
D-STAR ready
ID-51A VHF/UHF Dual Band Transceiver
D-STAR ready
• 5/2.5/1.0/0.5/0.1W Output • RX: 0.52–1.71, 88–174, 380-479 MHz** • AM/FM/FM-N/WFM/DV • 1304 Alphanumeric Memory Chls • Integrated GPS • D-STAR Repeater Directory • IPX7 Submersible
ID-31A UHF Digital Transceiver
5W Output Power • FM Analog Voice or D-STAR DV Mode • Built-in GPS Receiver • IPX7 Submersible • 1,252 Alphanumeric Memory Channels
D-STAR ready
ICOM
*Except 60M Band. **Frequency coverage may vary. Refer to owner's manual for exact specs. ***Tested to survive after being under 1m of water for 30 minutes. *¹Optional UX-9100 required. The Icom logo is a registered trademark of Icom Inc. 70037
CALL TOLL FREE
Phone Hours: 9:30 AM – 5:30 PM
Store Hours: 10:00 AM – 5:30 PM Closed Sun.
Toll free, incl. Hawaii, Alaska, Canada; call routed to nearest store; all HRO 800-lines can assist you, if the first line you call is busy, you may call another.
West............800-854-6046
Mountain......800-444-9476
Southeast......800-444-7927
Mid-Atlantic...800-444-4799
Northeast......800-644-4476
New England..800-444-0047
#1 in Customer Service
Come visit us online via the Internet at http://www.hamradio.com
AZ, CA, CO, GA, VA residents add sales tax. Prices, specifications, descriptions, subject to change without notice.
ANAHEIM, CA
(Near Disneyland)
933 N. Euclid St., 92801
(714) 533-7373
(800) 854-6046
Janet, KL7MF, Mgr.
anaheim@hamradio.com
BURBANK, CA
1525 W. Magnolia Bl., 91506
(818) 842-1786
(877) 892-1748
Eric, K6EJC, Mgr.
Magnolia between S. Victory & Buena Vista
burbank@hamradio.com
OAKLAND, CA
2210 Livingston St., 94606
(510) 534-5757
(877) 892-1745
Nick, AK6DX, Mgr.
I-880 at 23rd Ave. ramp
oakland@hamradio.com
SAN DIEGO, CA
5375 Kearny Villa Rd., 92123
(858) 560-4900
(877) 520-9623
Jerry, N5MCJ, Mgr.
Hwy. 163 & Claremont Mesa
sandiego@hamradio.com
SUNNYVALE, CA
510 Lawrence Exp. #102
94085
(408) 736-9496
(877) 892-1749
Jon, K6WV, Mgr.
So. from Hwy. 101
sunnyvale@hamradio.com
NEW CASTLE, DE
(Near Philadelphia)
1509 N. Dupont Hwy., 19720
(302) 322-7092
(800) 644-4476
Ken, N2OHD, Mgr.
RT.13 1/4 mi., So. I-295
delaware@hamradio.com
PORTLAND, OR
11705 S.W. Pacific Hwy.
97223
(503) 598-0555
(800) 765-4267
Leon, W7AD, Mgr.
Tigard-99W exit from Hwy. 5 & 217
portland@hamradio.com
DENVER, CO
8400 E. Iliff Ave. #9, 80231
(303) 745-7373
(800) 444-9476
John WØIG, Mgr.
denver@hamradio.com
PHOENIX, AZ
10613 N. 43rd Ave., 85029
(602) 242-3515
(800) 559-7388
Gary, N7GJ, Mgr.
Corner of 43rd Ave. & Peoria
phoenix@hamradio.com
ATLANTA, GA
6071 Buford Hwy., 30340
(770) 263-0700
(800) 444-7927
Mark, KJ4VO, Mgr.
Doraville, 1 mi. no. of I-285
atlanta@hamradio.com
WOODBRIDGE, VA
(Near Washington D.C.)
14803 Build America Dr.
22191
(703) 643-1063
(800) 444-4799
Steve, W4SHG, Mgr.
Exit 161, I-95, So. to US 1
virginia@hamradio.com
SALEM, NH
(Near Boston)
224 N. Broadway, 03079
(603) 898-3750
(800) 444-0047
Dave, N1EDU, Mgr.
Exit 1, I-93;
28 mi. No. of Boston
salem@hamradio.com

Quality Radio Equipment Since 1942
AMATEUR TRANSCEIVERS
Universal Radio is your authorized dealer for all major amateur radio equipment lines including: Alinco, Icom, Kenwood, Yaesu, TYT, Anytone, Wouxun and TDXone.
SHORTWAVE RECEIVERS
Whether your budget is $50, $500 or $5000, Universal can recommend a shortwave receiver to meet your needs. Universal is the only North American authorized dealer for all major shortwave lines including: Icom, Yaesu, Microtelecom, Ten-Tec, Grundig, etón, Sony, Sangean, Bonito, CommRadio, RFspace, Tecsun and Kaito.
WIDEBAND RECEIVERS
Universal Radio carries a broad selection of wideband and scanner receivers. Alinco, AOR, Icom, Yaesu, CommRadio, Whistler and Uniden Bearcat models are offered at competitive prices. Portable and tabletop models are available.
ANTENNAS AND ACCESSORIES
Your antenna is just as important as your radio. Universal carries antennas for every need, space requirement and budget. Please visit our website or request our catalog to learn more about antennas.
BOOKS
Whether your interest is amateur, shortwave or collecting; Universal has the very best selection of radio books and study materials.
■ USED EQUIPMENT
Universal carries an extensive selection of used amateur and shortwave equipment. Buy with confidence. All items have been tested by our service department and carry a 60 day warranty unless stated otherwise. We also buy radios. Request our printed used list or visit:
www.universal-radio.com
◆ VISIT OUR WEBSITE
Guaranteed lowest prices on the web? Not always. But we do guarantee that you will find the Universal website to be the most informative.
www.universal-radio.com
◆ VISIT OUR SHOWROOM
Showroom Hours
Mon.-Fri. 10:00 - 5:30
Saturday 10:00 - 3:00
270
Brice Rd.
Route 256
To Columbus
70
Exit 110A
Americana Pkwy.
6830
JC Penney
Brice Mall
Tussing Road
universal radio inc.
FREE 130 PAGE CATALOG
Our informative print catalog covers everything for the amateur, shortwave and scanner enthusiast. All items are also viewable at:
www.universal-radio.com
Universal Radio, Inc.
6830 Americana Pkwy.
Reynoldsburg, OH 43068
➤ 800 431-3939 Orders
➤ 614 866-4267 Information
➤ 614 866-2339 Fax
➤ dx@universal-radio.com
universal radio inc.
✓ Established in 1942.
✓ We ship worldwide.
✓ Visa, Master and Discover cards.
✓ Universal buys used radios.
✓ Returns subject to 15% restocking fee.

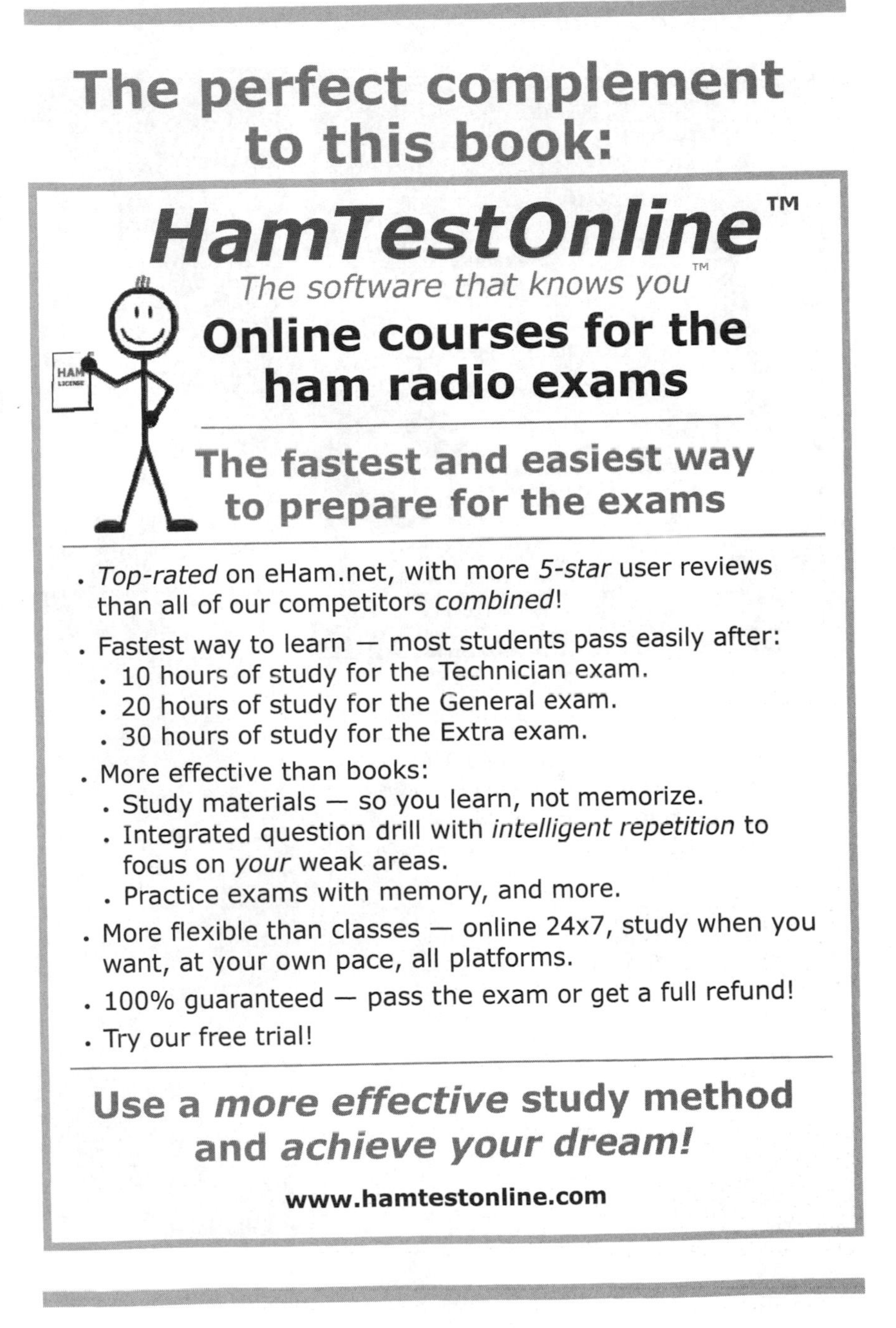
The perfect complement
to this book:
HamTestOnline™
The software that knows you™
HAM LICENSE
Online courses for the
ham radio exams
The fastest and easiest way
to prepare for the exams
. Top-rated on eHam.net, with more 5-star user reviews than all of our competitors combined!
. Fastest way to learn — most students pass easily after:
. 10 hours of study for the Technician exam.
. 20 hours of study for the General exam.
. 30 hours of study for the Extra exam.
. More effective than books:
. Study materials — so you learn, not memorize.
. Integrated question drill with intelligent repetition to focus on your weak areas.
. Practice exams with memory, and more.
. More flexible than classes — online 24x7, study when you want, at your own pace, all platforms.
. 100% guaranteed — pass the exam or get a full refund!
. Try our free trial!
Use a more effective study method
and achieve your dream!
www.hamtestonline.com

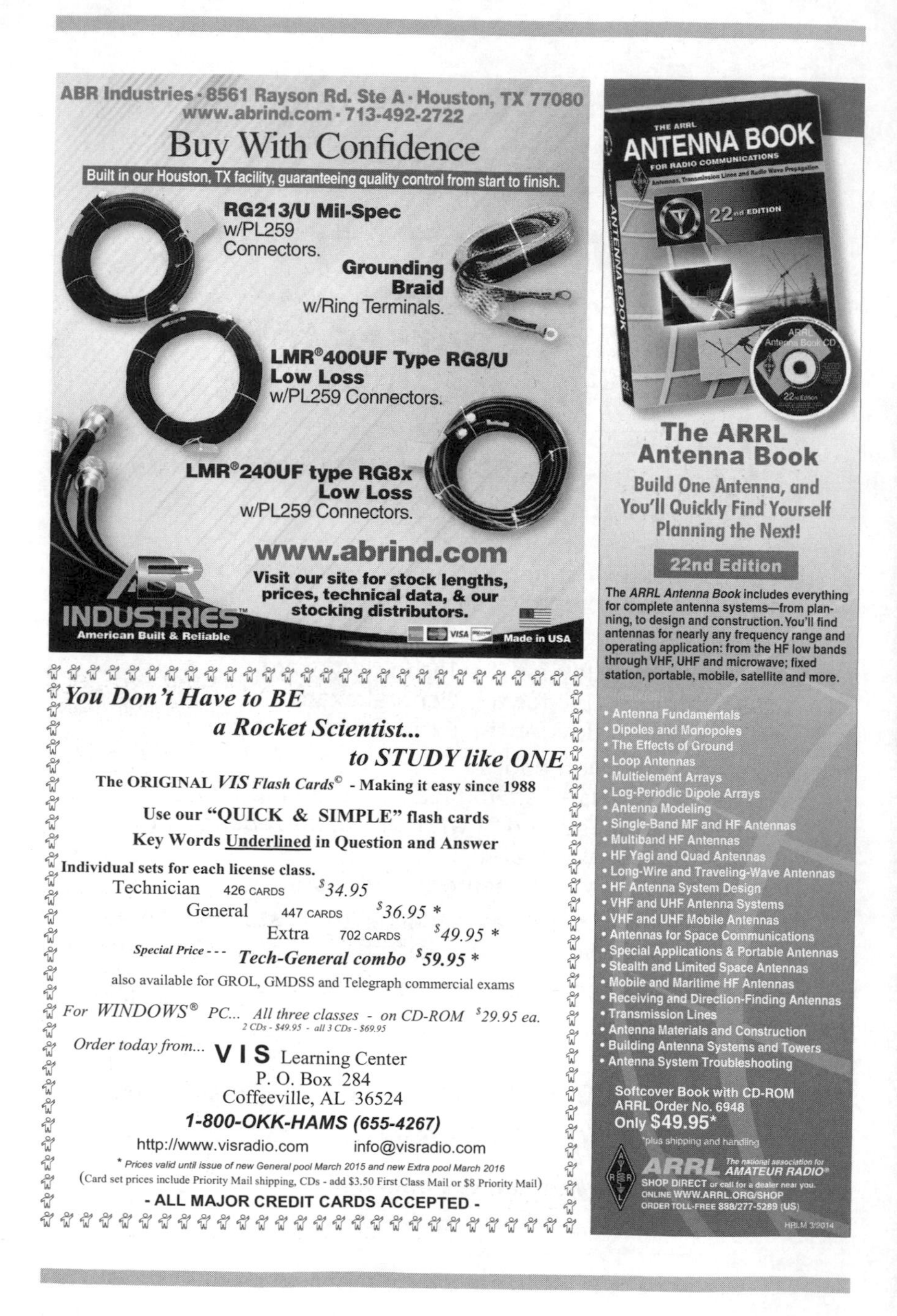
ABR Industries · 8561 Rayson Rd. Ste A · Houston, TX 77080
www.abrind.com · 713-492-2722
Buy With Confidence
Built in our Houston, TX facility, guaranteeing quality control from start to finish.
RG213/U Mil-Spec
w/PL259
Connectors.
Grounding
Braid
w/Ring Terminals.
LMR®400UF Type RG8/U
Low Loss
w/PL259 Connectors.
LMR®240UF type RG8x
Low Loss
w/PL259 Connectors.
www.abrind.com
Visit our site for stock lengths, prices, technical data, & our stocking distributors.
INDUSTRIES
American Built & Reliable
Made in USA
You Don't Have to BE
a Rocket Scientist...
to STUDY like ONE
The ORIGINAL VIS Flash Cards© - Making it easy since 1988
Use our "QUICK & SIMPLE" flash cards
Key Words Underlined in Question and Answer
Individual sets for each license class.
Technician 426 CARDS $34.95
General 447 CARDS $36.95 *
Extra 702 CARDS $49.95 *
Special Price --- Tech-General combo $59.95 *
also available for GROL, GMDSS and Telegraph commercial exams
For WINDOWS® PC... All three classes - on CD-ROM $29.95 ea.
2 CDs - $49.95 - all 3 CDs - $69.95
Order today from... VIS Learning Center
P. O. Box 284
Coffeeville, AL 36524
1-800-OKK-HAMS (655-4267)
http://www.visradio.com info@visradio.com
* Prices valid until issue of new General pool March 2015 and new Extra pool March 2016
(Card set prices include Priority Mail shipping, CDs - add $3.50 First Class Mail or $8 Priority Mail)
- ALL MAJOR CREDIT CARDS ACCEPTED -
THE ARRL ANTENNA BOOK FOR RADIO COMMUNICATIONS
22nd EDITION
The ARRL
Antenna Book
Build One Antenna, and You'll Quickly Find Yourself Planning the Next!
22nd Edition
The ARRL Antenna Book includes everything for complete antenna systems—from planning, to design and construction. You'll find antennas for nearly any frequency range and operating application: from the HF low bands through VHF, UHF and microwave; fixed station, portable, mobile, satellite and more.
• Antenna Fundamentals
• Dipoles and Monopoles
• The Effects of Ground
• Loop Antennas
• Multielement Arrays
• Log-Periodic Dipole Arrays
• Antenna Modeling
• Single-Band MF and HF Antennas
• Multiband HF Antennas
• HF Yagi and Quad Antennas
• Long-Wire and Traveling-Wave Antennas
• HF Antenna System Design
• VHF and UHF Antenna Systems
• VHF and UHF Mobile Antennas
• Antennas for Space Communications
• Special Applications & Portable Antennas
• Stealth and Limited Space Antennas
• Mobile and Maritime HF Antennas
• Receiving and Direction-Finding Antennas
• Transmission Lines
• Antenna Materials and Construction
• Building Antenna Systems and Towers
• Antenna System Troubleshooting
Softcover Book with CD-ROM
ARRL Order No. 6948
Only $49.95*
*plus shipping and handling
ARRL The national association for AMATEUR RADIO®
SHOP DIRECT or call for a dealer near you.
ONLINE WWW.ARRL.ORG/SHOP
ORDER TOLL-FREE 888/277-5289 (US)
HRLM 3/2014

Heavy-Duty FM Dual Band Mobile with
Exceptionally Wide Receiver Coverage*
*108 to 520 MHz/ 700 to 999.99 MHz (Cellular blocked)
Large Backlit LCD Display for easy operation
Stable RF Power (50 Watts VHF / 45 Watts UHF)
Reliable performance in harsh environments
5 ppm Frequency Stability (-4° F to +140° F)
1000 Memory Channels for serious users
Yaesu Unique Power Saving Circuit Design Minimizes Vehicle Battery Drain
•Separation Kit for Remote Mounting (optional separation kit YSK-7800 requires)
The King of Mobile
Massive Heatsink guarantees 75 Watts of Solid RF Power with No Cooling Fan Needed
Loud 3 Watts of Audio Output for noisy environments
Large 6 Digit Backlit LCD for excellent visibility
200 Memory Channels for serious users
75 WATTS
50 W 2 m/70 cm* DUAL BAND FM TRANSCEIVER
*70 cm 45 W
FT-7900R
Size: 5.5" (W) x 1.6" (H) x 6.6" (D) / Weight: 2.2 lb
2 m/70 cm DUAL BAND
2 m MONO BAND
HEAVY-DUTY 75 W 2 m FM TRANSCEIVER
FT-2900R
Size: 6.3" (W) x 2.0" (H) x 7.3" (D) / Weight: 4.0 lb
55 WATTS
Best Selling, Reliable Mobile
55 Watts of Solid RF Power within a compact footprint
Loud 3 Watts of Audio Output Power for noisy enviroments
Large 6 Digit Backlit LCD for excellent visibility
200 Memory Channels for serious users
ULTRA RUGGED 55 W 2 m FM TRANSCEIVER
FT-1900R
2 m MONO BAND
Size: 5.5" (W) x 1.6" (H) x 5.8" (D) / Weight: 2.2 lb
YAESU
The radio
For the latest Yaesu news, visit us on the Internet: http://www.yaesu.com
Specifications subject to change without notice. Some accessories and/or options may be standard in certain areas. Frequency coverage may differ in some countries. Check with your local Yaesu Dealer for specific details.
YAESU USA
6125 Phyllis Drive,
Cypress, CA 90630 (714) 827-7600

Field Gear That Goes The Distance!
FT-897D
HF/VHF/UHF Portable Operation Powerful Transceiver
The Ultimate Emergency Communications Radio
Rugged, Innovative Multi-Band
Operates on the SSB, CW, AM, FM, and Digital Modes
Wide Frequency Coverage
20-Watt Portable Operation Using Internal Batteries
100 Watts When Using an External 13.8-Volt DC Power Source
FT-817ND
The Ultimate Backpack, Multi-Mode Portable Transceiver
Self-Contained
Battery-Powered
Covering the HF, VHF, and UHF Bands
Provides up to Five Watts of Power Output
SSB, CW, AM, FM, Packet, or SSB-based Digital Modes like PSK31
FT-857D
The World's Smallest HF/VHF/UHF Mobile Transceiver
Ultra-Compact Package
Ideal for Mobile or External Battery Portable Work
Wide Frequency Coverage
Optional Remote-Head
High-Performance Mobile Operation
FT-450D
HF/50 MHz 100 W Easy to Operate All Mode Transceiver
Illuminated Key Buttons
300Hz / 500Hz / 2.4 kHz CW IF Filter
Foot Stand
Classically Designed Main Dial and Knobs
Dynamic Microphone MH-31 A8J Included
YAESU
The radio
YAESU USA
6125 Phyllis Drive, Cypress, CA 90630 Phone: (714) 827-7600
http://www.yaesu.com

About the ARRL

The seed for Amateur Radio was planted in the 1890s, when Guglielmo Marconi began his experiments in wireless telegraphy. Soon he was joined by dozens, then hundreds, of others who were enthusiastic about sending and receiving messages through the air—some with a commercial interest, but others solely out of a love for this new communications medium. The United States government began licensing Amateur Radio operators in 1912.

By 1914, there were thousands of Amateur Radio operators—hams—in the United States. Hiram Percy Maxim, a leading Hartford, Connecticut inventor and industrialist, saw the need for an organization to band together this fledgling group of radio experimenters. In May 1914 he founded the American Radio Relay League (ARRL) to meet that need.

Today ARRL, with approximately 150,000 members, is the largest organization of radio amateurs in the United States. The ARRL is a not-for-profit organization that:

- promotes interest in Amateur Radio communications and experimentation
- represents US radio amateurs in legislative matters, and
- maintains fraternalism and a high standard of conduct among Amateur Radio operators.

At ARRL headquarters in the Hartford suburb of Newington, the staff helps serve the needs of members. ARRL is also International Secretariat for the International Amateur Radio Union, which is made up of similar societies in 150 countries around the world.

ARRL publishes the monthly journal *QST*, as well as newsletters and many publications covering all aspects of Amateur Radio. Its headquarters station, W1AW, transmits bulletins of interest to radio amateurs and Morse code practice sessions. The ARRL also coordinates an extensive field organization, which includes volunteers who provide technical information and other support services for radio amateurs as well as communications for public-service activities. In addition, ARRL represents US amateurs with the Federal Communications Commission and other government agencies in the US and abroad.

Membership in ARRL means much more than receiving *QST* each month. In addition to the services already described, ARRL offers membership services on a personal level, such as the ARRL Volunteer Examiner Coordinator Program and a QSL bureau.

Full ARRL membership (available only to licensed radio amateurs) gives you a voice in how the affairs of the organization are governed. ARRL policy is set by a Board of Directors (one from each of 15 Divisions) elected by the full members they represent. The day-to-day operation of ARRL HQ is managed by a Chief Executive Officer and his staff.

No matter what aspect of Amateur Radio attracts you, ARRL membership is relevant and important. There would be no Amateur Radio as we know it today were it not for the ARRL. We would be happy to welcome you as a member! (An Amateur Radio license is not required for Associate Membership.) For more information about ARRL and answers to any questions you may have about Amateur Radio, write or call:

ARRL—The national association for Amateur Radio
225 Main Street
Newington CT 06111-1494

Voice: 860-594-0200
Fax: 860-594-0259

E-mail: **hq@arrl.org**
Internet: **www.arrl.org/**

Prospective new amateurs call (toll-free):
800-32-NEW HAM (800-326-3942)
You can also contact us via e-mail at **newham@arrl.org**
or check out **ARRLWeb** at **www.arrl.org/**

Notes

Notes

Notes

Notes

Notes

Notes

We're interested in hearing your comments on this book and what you'd like to see in future editions. Please email comments to us at **pubsfdbk@arrl.org**, including your name, call sign, email address and the title, edition and printing of this book. Or you can copy this form, fill it out, and mail to ARRL, 225 Main St, Newington, CT 06111-1494.

Please check the box that best answers these questions:

How well did this book prepare you for your exam?

☐ Very Well ☐ Fairly Well ☐ Not Very Well

Did you pass? ☐ Yes ☐ No

Where did you purchase this book?

☐ From ARRL directly ☐ From an ARRL dealer

Is there a dealer who carries ARRL publications within:

☐ 5 miles ☐ 15 miles ☐ 30 miles of your location? ☐ Not sure.

If licensed, what is your license class? ______________________

Name ______________________ ARRL member? ☐ Yes ☐ No

Call Sign ____________

Address ______________________

City, State/Province, ZIP/Postal Code ______________________

Daytime Phone ( ) ____________ Age ______ E-mail ____________

If licensed, how long? ______________________

Other hobbies ______________________

____

Occupation ______________________

For ARRL use only	**Tech Q&A**
Edition	6 7 8 9 10 11
Printing	2 3 4 5 6 7 8 9 10 11